Compendium of Plant Genomes

Series editor

Chittaranjan Kole, Raja Ramanna Fellow, Department of Atomic Energy, Government of India, Kalyani, India

Whole-genome sequencing is at the cutting edge of life sciences in the new millennium. Since the first genome sequencing of the model plant Arabidopsis thaliana in 2000, whole genomes of about 70 plant species have been sequenced and genome sequences of several other plants are in the pipeline. Research publications on these genome initiatives are scattered on dedicated web sites and in journals with all too brief descriptions. The individual volumes elucidate the background history of the national and international genome initiatives; public and private partners involved; strategies and genomic resources and tools utilized; enumeration on the sequences and their assembly; repetitive sequences; gene annotation and genome duplication. In addition, synteny with other sequences, comparison of gene families and most importantly potential of the genome sequence information for gene pool characterization and genetic improvement of crop plants are described.

Interested in editing a volume on a crop or model plant? Please contact Dr. Kole, Series Editor, at ckole2012@gmail.com

More information about this series at http://www.springer.com/series/11805

Rajeev K. Varshney · Rachit K. Saxena
Scott A. Jackson
Editors

The Pigeonpea Genome

Springer

Editors
Rajeev K. Varshney
Center of Excellence in Genomics (CEG)
International Crops Research Institute for the Semi-Arid Tropics (ICRISAT)
Patancheru
India

Scott A. Jackson
Center for Applied Genetic Technologies
University of Georgia
Athens, Georgia
USA

Rachit K. Saxena
Center of Excellence in Genomics (CEG)
International Crops Research Institute for the Semi-Arid Tropics (ICRISAT)
Patancheru
India

ISSN 2199-4781 ISSN 2199-479X (electronic)
Compendium of Plant Genomes
ISBN 978-3-319-87637-5 ISBN 978-3-319-63797-6 (eBook)
DOI 10.1007/978-3-319-63797-6

Softcover re-print of the Hardcover 1st edition 2017

Printed on acid-free paper

This Springer imprint is published by Springer Nature
The registered company is Springer International Publishing AG
The registered company address is: Gewerbestrasse 11, 6330 Cham, Switzerland

This book series is dedicated to
my wife Phullara, and our children
Sourav, and Devleena

Chittaranjan Kole

Preface to the Series

Genome sequencing has emerged as the leading discipline in the plant sciences coinciding with the start of the new century. For much of the twentieth century, plant geneticists were only successful in delineating putative chromosomal location, function, and changes in genes indirectly through the use of a number of 'markers' physically linked to them. These included visible or morphological, cytological, protein, and molecular or DNA markers. Among them, the first DNA marker, the RFLPs, introduced a revolutionary change in plant genetics and breeding in the mid-1980s, mainly because of their infinite number and thus potential to cover maximum chromosomal regions, phenotypic neutrality, absence of epistasis, and codominant nature. An array of other hybridization-based markers, PCR-based markers, and markers based on both facilitated construction of genetic linkage maps, mapping of genes controlling simply inherited traits, and even gene clusters (QTLs) controlling polygenic traits in a large number of model and crop plants. During this period, a number of new mapping populations beyond F2 were utilized and a number of computer programs were developed for map construction, mapping of genes, and for mapping of polygenic clusters or QTLs. Molecular markers were also used in studies of evolution and phylogenetic relationship, genetic diversity, DNA-fingerprinting, and map-based cloning. Markers tightly linked to the genes were used in crop improvement employing the so-called marker-assisted selection. These strategies of molecular genetic mapping and molecular breeding made a spectacular impact during the last one and a half decades of the twentieth century. But still they remained 'indirect' approaches for elucidation and utilization of plant genomes since much of the chromosomes remained unknown and the complete chemical depiction of them was yet to be unraveled.

Physical mapping of genomes was the obvious consequence that facilitated development of the 'genomic resources' including BAC and YAC libraries to develop physical maps in some plant genomes. Subsequently, integrated genetic–physical maps were also developed in many plants. This led to the concept of structural genomics. Later on, emphasis was laid on EST and transcriptome analysis to decipher the function of the active gene sequences leading to another concept defined as functional genomics. The advent of techniques of bacteriophage gene and DNA sequencing in the 1970s was extended to facilitate sequencing of these genomic resources in the last decade of the twentieth century.

As expected, sequencing of chromosomal regions would have led to too much data to store, characterize, and utilize with the-then available computer software could handle. But development of information technology made the life of biologists easier by leading to a swift and sweet marriage of biology and informatics, and a new subject was born—bioinformatics.

Thus, evolution of the concepts, strategies, and tools of sequencing and bioinformatics reinforced the subject of genomics—structural and functional. Today, genome sequencing has traveled much beyond biology and involves biophysics, biochemistry, and bioinformatics!

Thanks to the efforts of both public and private agencies, genome sequencing strategies are evolving very fast, leading to cheaper, quicker, and automated techniques right from clone-by-clone and whole-genome shotgun approaches to a succession of second generation sequencing methods. Development of software of different generations facilitated this genome sequencing. At the same time, newer concepts and strategies were emerging to handle sequencing of the complex genomes, particularly the polyploids.

It became a reality to chemically—and so directly—define plant genomes, popularly called whole-genome sequencing or simply genome sequencing.

The history of plant genome sequencing will always cite the sequencing of the genome of the model plant Arabidopsis thaliana in 2000 that was followed by sequencing the genome of the crop and model plant rice in 2002. Since then, the number of sequenced genomes of higher plants has been increasing exponentially, mainly due to the development of cheaper and quicker genomic techniques and, most importantly, development of collaborative platforms such as national and international consortia involving partners from public and/or private agencies.

As I write this preface for the first volume of the new series 'Compendium of Plant Genomes,' a net search tells me that complete or nearly complete whole-genome sequencing of 45 crop plants, eight crop and model plants, eight model plants, 15 crop progenitors and relatives, and three basal plants is accomplished, the majority of which are in the public domain. This means that we nowadays know many of our model and crop plants chemically, i.e., directly, and we may depict them and utilize them precisely better than ever. Genome sequencing has covered all groups of crop plants. Hence, information on the precise depiction of plant genomes and the scope of their utilization is growing rapidly every day. However, the information is scattered in research articles and review papers in journals and dedicated Web pages of the consortia and databases. There is no compilation of plant genomes and the opportunity of using the information in sequence-assisted breeding or further genomic studies. This is the underlying rationale for starting this book series, with each volume dedicated to a particular plant.

Plant genome science has emerged as an important subject in academia, and the present compendium of plant genomes will be highly useful both to students and teaching faculties. Most importantly, research scientists involved in genomics research will have access to systematic deliberations on the plant genomes of their interest. Elucidation of plant genomes is of interest not only for the geneticists and breeders, but also for practitioners of an array of plant science disciplines, such as taxonomy, evolution, cytology, physiology, pathology, entomology, nematology, crop production, biochemistry, and obviously bioinformatics. It must be mentioned that information regarding each plant genome is ever-growing. The contents of the volumes of this compendium are therefore focusing on the basic aspects of the genomes and their utility. They include information on the academic and/ or economic importance of the plants, description of their genomes from a molecular genetic and cytogenetic point of view, and the genomic resources developed. Detailed deliberations focus on the background history of the national and international genome initiatives, public and private partners involved, strategies and genomic resources and tools utilized, enumeration on the sequences and their assembly, repetitive sequences, gene annotation, and genome duplication. In addition, synteny with other sequences, comparison of gene families, and, most importantly, potential of the genome sequence information for gene pool characterization through genotyping by sequencing (GBS) and genetic improvement of crop plants have been described. As expected, there is a lot of variation of these topics in the volumes based on the information available on the crop, model, or reference plants.

I must confess that as the series editor, it has been a daunting task for me to work on such a huge and broad knowledge base that spans so many diverse plant species. However, pioneering scientists with lifetime experience and expertise on the particular crops did excellent jobs editing the respective volumes. I myself have been a small science worker on plant genomes since the mid-1980s and that provided me the opportunity to personally know several stalwarts of plant genomics from all over the globe. Most, if not all, of the volume editors are my longtime friends and colleagues. It has been highly comfortable and enriching for me to work with them on this book series. To be honest, while working on this series I have been and will remain a student first, a science worker second, and a series editor last. And I must express my gratitude to the volume editors and the chapter authors for providing me the opportunity to work with them on this compendium.

I also wish to mention here my thanks and gratitude to the Springer staff, Dr. Christina Eckey and Dr. Jutta Lindenborn in particular, for all their constant and cordial support right from the inception of the idea.

I always had to set aside additional hours to edit books besides my professional and personal commitments—hours I could and should have given to my wife, Phullara, and our kids, Sourav, and Devleena. I must mention that they not only allowed me the freedom to take away those hours from them but also offered their support in the editing job itself. I am really not sure whether my dedication of this compendium to them will suffice to do justice to their sacrifices for the interest of science and the science community.

Kalyani, India Chittaranjan Kole

Preface to the Volume

The order of four nucleotides, i.e., adenine, cytosine, guanine, and thymine constitutes an organism's DNA or genome. Genome sequencing is to understand the order of nucleotides, or bases, in a genome which is responsible for basic behavior of every organism. With the current advances in the genome sequencing, the entire genomes have been decoded in a number of organisms. This has provided ways to manipulate genome constitutions for achieving desirable phenotypes in crops species as well. Such efforts in pigeonpea (*Cajanus cajan*) have been slow or negligible until onset of twenty-first century. The availability of reference genome sequence for cultivated pigeonpea has accelerated understanding the basic biology, understanding and deploying modern approaches for candidate gene discovery, and marker development for key traits.

Pigeonpea is an important legume crop in arid and semi-arid regions of the world. It provides nutritional food to the vegetarian families living in the various countries of the world. The early cultivated pigeonpea has been domesticated from its wild progenitor species, i.e., *C. cajanifolius* in central India around 3,000 years ago. This crop has unique feature of often cross-pollination behavior, which has been used in developing high yielding hybrids. Beginning of second decade of twenty-first century has been a starting phase of revolutionizing pigeonpea research as a number of high yielding hybrids have been released and reference genome sequence became available for candidate gene discovery, high-resolution trait mapping, and marker development and genomics-assisted breeding.

This book is well-timed in pigeonpea research as part of the genome compendium series for different crops. It contains 10 different chapters providing detailed overview on different aspects of botanical classification, genetics, genomics, and breeding of pigeonpea. This book not only provides information on recent advances on genome sequencing, genome architecture, genetic mapping, and marker identification but also presents future guidelines of research by deploying modern genomics tools in conjunction with breeding.

A total of 26 authors have contributed 10 chapters for this volume (see Appendix I). The editors of this volume are grateful to all the authors for their contribution in writing chapters and reviewers (see Appendix II) for their constructive suggestions and corrections helping in improving the quality of the chapters further. The editors are also thankful to Dr. David

Bergvinson, Director General, ICRISAT and Dr. Peter Carberry, Deputy Director General—Research, ICRISAT for their support. The editors thank Prof. C. Kole, Series Editor for his invitation and help in editing this volume. The cooperation received from Abirami Purushothaman, Jegadeeswari Diravidamani, Naresh Kumar and Jutta Lindenborn from Springer has been a great help in completion of this book and is gratefully acknowledged.

In addition to above, we also appreciate and recognize cooperation and moral support from our family members for sparing us precious time for editorial work that we should have spent with our respective families. RKV acknowledges the help and support of wife (Monika), son (Prakhar), and daughter (Preksha) who allowed their time to be taken away to fulfill RKV's editorial responsibilities in addition to research and other administrative duties at ICRISAT. Similarly, RKS is grateful to his wife (Shelly) and two young sons (Aniruddha and Madhav) for their help and moral support in doing editorial responsibilities in addition to research duties at ICRISAT. SJ also acknowledges his wife (Julie) for support.

Editors hope that their efforts in compiling the information on different aspects of pigeonpea will help the pigeonpea genomics and breeding researchers in developing better understanding and research strategies. This book will also benefit students, academicians, and policy makers in updating their knowledge on recent advances in pigeonpea research.

ICRISAT, India — Rajeev K. Varshney
ICRISAT, India — Rachit K. Saxena
UGA, USA — Scott A. Jackson

Contents

Contributors

Mohd Ashraf Bhat Division of Plant Breeding and Genetics, Sher-e-Kashmir University of Agricultural Sciences & Technology of Kashmir (SKUAST-K), Kashmir, India

Abhishek Bohra Crop Improvement Division, Indian Institute of Pulses Research (IIPR), Kanpur, India

M. Byre Gowda All India Co-ordinated Research Project on Pigeonpea, University of Agricultural Sciences, GKVK, Bengaluru, Karnataka, India

Scott A. Jackson The University of Georgia, College of Agricultural and Environmental Sciences, Athens, GA, USA

Rintu Jha Crop Improvement Division, Indian Institute of Pulses Research (IIPR), Kanpur, India

Rohit Joshi Plant Stress Biology, International Centre for Genetic Engineering and Biotechnology, New Delhi, India

Mayank Kaashyap School of Applied Sciences, Health Innovations Research Institute, RMIT University, Melbourne, VIC, Australia

Mulualem T. Kassa National Research Council, Saskatoon, SK, Canada; Agriculture and Agri-Food Canada, Morden Research and Development Centre, Morden, MB, Canada

Christopher P. Krieg Department of Biological Sciences and International Center for Tropical Botany, Florida International University, Miami, FL, USA; Department of Botany, University of Florida, Gainesville, FL, USA

K.B. Saxena Villa 17, NMC Housing, Al Ain, Abu Dhabi, United Arab Emirates

Nalini Mallikarjuna International Crops Research Institute for the Semi-Arid Tropics, Patancheru, India; , Mysore, Karnataka, India

Reyazul Rouf Mir Division of Plant Breeding and Genetics, Sher-e-Kashmir University of Agricultural Sciences & Technology of Kashmir (SKUAST-K), Kashmir, India

Raj K. Mishra Crop Improvement Division, Indian Institute of Pulses Research (IIPR), Kanpur, India

Nidhi Mohan ANGRAU, Guntur, Andhra Pradesh, India

Gaurav Pandey Crop Improvement Division, Indian Institute of Pulses Research (IIPR), Kanpur, India

Shalini Pareek Crop Improvement Division, Indian Institute of Pulses Research (IIPR), Kanpur, India

G.A. Parray Division of Plant Breeding and Genetics, Sher-e-Kashmir University of Agricultural Sciences & Technology of Kashmir (SKUAST-K), Kashmir, India

Irshad Ahmad Rather Division of Plant Breeding and Genetics, Sher-e-Kashmir University of Agricultural Sciences & Technology of Kashmir (SKUAST-K), Kashmir, India

C.V. Sameer Kumar Pigeonpea Breeding, ICRISAT, Patancheru, Telangana, India

S.J. Satheesh Naik Pigeonpea Breeding, ICAR-IIPR, Kanpur, UP, India

Rachit K. Saxena Pigeonpea Breeding, International Crops Research Institute for the Semi-Arid Tropics (ICRISAT), Patancheru, India

Farindra Singh Crop Improvement Division, Indian Institute of Pulses Research (IIPR), Kanpur, India

Indra P. Singh Crop Improvement Division, Indian Institute of Pulses Research (IIPR), Kanpur, India

Vikas K. Singh International Crops Research Institute for the Semi-Arid Tropics, Patancheru, India

Rajeev K. Varshney Pigeonpea Breeding, Centre of Excellence in Genomics (CEG), International Crops Research Institute for the Semi-Arid Tropics, Patancheru, India

Eric J.B. von Wettberg Department of Biological Sciences and International Center for Tropical Botany, Florida International University, Miami, FL, USA; Department of Plant and Soil Science, University of Vermont, Burlington, VT, USA

Reviewers

Bohra, Abhishek, Indian Institute of Pulses Research (IIPR), Kanpur, India

Bhuvaneshwaragouda, R. Patil, University of Agricultural Science's (UAS-D), Dharwad, India

Saxena, KB, International Crops Research Institute for the Semi-Arid Tropics (ICRISAT), Patancheru, India

Singh, Indra Prakash, Indian Institute of Pulses Research (IIPR), Kanpur, India

Singh, Narendra Pratap, Indian Institute of Pulses Research (IIPR), Kanpur, India

Sharma, Shivali, International Crops Research Institute for the Semi-Arid Tropics (ICRISAT), Patancheru, India

Malik, Sundip, G.B. Pant University Argiculture & Technolgy, U.S. Nagar, India

Silim, N. Said, International Center for Agricultural Research in the Dry Areas (ICARDA), Addis Ababa, Ethiopia

The Pigeonpea Genome: An Overview

1

Rajeev K. Varshney, Rachit K. Saxena and Scott A. Jackson

Abstract

First two decades of twenty-first century have witnessed a number of advances in genetics and genomics research of pigeonpea. These advances have enhanced our understanding of structural and functional aspects of genome and also provided us opportunities to deal with constraints impeding production of pigeonpea in precise and faster manner. Availability of the draft genome sequence and large-scale molecular markers has made it possible to map traits of interest in speedy manner. Although germplasm re-sequencing has already been started in pigeonpea, large-scale germplasm including elite breeding line, landraces and wild species is expected to be fully sequenced very soon. These sequencing efforts coupled with functional genomics and systems biology will facilitate the identification of genes/gene networks that are involved in expression of agronomically valuable traits. For accelerating genetic gains in the crop breeding, selection efficiency needs to be enhanced by integrating modern genomics in breeding efforts. This book provides a critical assessment on current status as well as future prospects on different aspects of genome, trait mapping, germplasm research and genomics-assisted breeding. This chapter introduces and provides highlights of different chapters of the book.

1.1 Introduction

Pigeonpea is considered a resilient crop to environmental constraints and climate changes. It is an important crop for providing proteins in food to the poor people especially in Asia. In present-day scenario, agricultural focus is shifting to "nutritional food security" from just "food security". Therefore, pulse crops such as

R.K. Varshney (✉) · R.K. Saxena
International Crops Research Institute for the Semi-Arid Tropics (ICRISAT), Patancheru 502324, India

S.A. Jackson
The University of Georgia, College of Agricultural and Environmental Sciences, Athens, GA 30602, USA

R.K. Varshney et al. (eds.), *The Pigeonpea Genome*,
Compendium of Plant Genomes, DOI 10.1007/978-3-319-63797-6_1

pigeonpea have potential for sustainable agriculture to meet the needs of a fast-growing human population worldwide. However, pigeonpea suffers with a low yield levels due to the exposure of the crop to a number of biotic and abiotic stresses (Saxena et al. 2014). On the other hand, lack of systematic public funding and minimal or no industrial funding support to pigeonpea research and development have also contributed to slow development of varieties with the limited genetic gains in the past. As a result, the crop productivity has remained stagnant for about last six decades (Fig. 1.1). Emphasis of research and funding and support of private sector in pigeonpea was also highlighted by Mr. Bill Gates during his visit to ICRISAT in his blog entitled "Making a Better Pigeonpea" as following: "*It's also an example of what agricultural development people call an orphan crop, a crop that's important to many of the world's poorest people yet largely ignored by the big agriculture companies. Those companies focus on high-value crops like corn and soy that are building blocks of rich-world diets and industry. As a result, their research has boosted yields of those crops by making them more resistant to insects, disease, and drought. The orphans haven't seen many, if any, of those kinds of improvements*" (https://www.gatesnotes.com/Development/Visiting-ICRISAT-Agricultural-Research-Center).

Genomics has made positive interventions on enhancing yield and developing better varieties in many cereals and few legume crops (Varshney et al. 2011). However, the pigeonpea crop, until recently, has largely been untouched with this genomic revolution. As a result, pigeonpea has often referred as "orphan crop". However, due to the decline in sequencing cost and strong partnerships across different continents, draft genome was assembled for pigeonpea (Varshney et al. 2012) that has provided tools and opportunities for pigeonpea improvement. These tools can be used in a number of ways such as deployment of early generation screening of large segregating populations for must-have traits, gene pyramiding for multiple resistance to specific pathogens and pests within the same cultivar, introgression of superior alleles from landraces and wild species in cultivated material.

In recent years, significant progress has been achieved in the area of pigeonpea genomics. For instance, large-scale genomic and transcriptome resources have been developed and used in germplasm research, trait mapping, molecular breeding as well as functional genomics research (Pazhamala et al. 2015). The present book entitled "The Pigeonpea Genome" therefore aims to present available information on different aspects of research in the area of germplasm, genetics, genomics and breeding. The book also provides

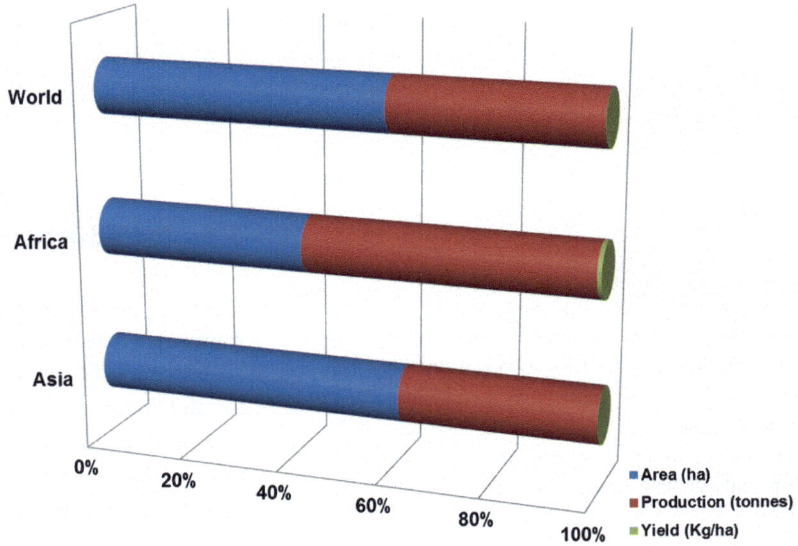

Fig. 1.1 Pigeonpea area, production and productivity in different regions during 2014

approaches and strategies to apply in breeding. Furthermore, in addition to achievements, constraints and future prospects of applications of modern genomics in pigeonpea improvement have also been appraised. The introductory chapter provides an overview of all the chapters of the book in following sections.

1.2 Crop Characteristics, Botanical Description and Wide Crossing

Chapter 2 entitled "Key Plant and Grain Characteristics and Their Importance in Breeding and Adaptation of Pigeonpea Cultivars" by KB Saxena and colleagues provides detailed information on the potential role of different plant and grain characteristics in determining yield and stability of pigeonpea. Yield enhancement is an ultimate target for any crop improvement program, however; there is no straight route for breeding high-yielding cultivars. Breeders have relied on selecting individual traits, which contributing to yield directly or indirectly and limited success has been achieved in pigeonpea. Therefore, in Chap. 2, authors have highlighted various qualitative and quantitative traits related to seed yield, quality and those preferred for marketing and milling. A brief description about their inheritance and association with yield in pigeonpea has also been provided.

Chapter 3, Botanical Description of Pigeonpea authored by Sameer Kumar et al., provides information on the genetic structure of the genus including its origin, variability and geographical distribution of various species. The detailed description on pollination behaviour, adaptability to a range of soil types, temperature and rainfall, ability of nitrogen fixation, root system, branching pattern, growth habit or plant architecture has been provided in this chapter.

Nalini Mallikarjuna and colleagues provide detailed information on the importance and utilization of rich source of genetic variations in Chap. 4 entitled Wide Crossing Technology for Pigeonpea Improvement. This chapter also summarized a thorough knowledge of crossability and concerted efforts to effectively utilize the immense variation present in the secondary, tertiary and quaternary gene pool. In the end of this chapter, an emphasis has been given to use advances in genomics wide crossing program.

1.3 Genetic Resources and Trait Mapping

Bohra et al. provide information on the modern genomic tools for pigeonpea improvement in Chap. 5. This chapter reviews the progress on generation of genomic resources and highlights their importance in designing future crop breeding schemes. This chapter has assembled the information on the collaborative research efforts which have facilitated development of genomic tools (mapping populations, molecular markers, genome sequence, transcriptome, etc.) for pigeonpea improvement during the last ten years (Pazhamala et al. 2015). Subsequently, Irshad Ahmad Rather and colleagues in Chap. 6 compile molecular mapping efforts on genes and quantitative trait loci (QTL) in pigeonpea. Chapter 7 entitled "Germplasm Characterization and Trait Discovery" authored by Christopher P Krieg and colleagues discusses development of core and minicore collections for better utilization of diverse germplasm in routine pigeonpea breeding program. The chapter also presents about the research priorities for important traits such as yield, resistance to biotic and abiotic stress. Further deployment of modern genomics approaches has been suggested for accelerated trait/gene discovery and development of appropriate genomics tools for key traits to deploy them in routine breeding program to utilize germplasm collection.

1.4 Genome Sequence and Beyond

Two chapters, i.e. Chaps. 8 and 9, explain about various efforts undertaken for *de novo* genome sequencing of pigeonpea and provide background history on genome sequencing efforts and current status on deployment of genome sequence information for crop improvement.

These chapters related to genome sequencing have also discussed about recent advances in establishing high-density genotyping platforms such as genotyping by sequencing (GBS) and Axiom®*CajanusSNP* Array*: 56 K*, whole genome re-sequencing efforts, etc. In the last chapter (Chap. 10) entitled "Future Prospects", the editors of book have presented a concise view on future prospects of pigeonpea research. Next-generation breeding including use of next-generation sequencing and high-throughput genotyping for early generation screening, marker-assisted selection, marker-assisted backcrossing and genomic selection as well as genome editing coupled with other advances has been proposed to achieve rapid genetic gains for pigeonpea improvement.

1.5 Conclusion

This book provides up-to-date information about pigeonpea genome and its utilization for germplasm research, advancing genetics, genomics and accelerating breeding practices by a panel of lead pigeonpea scientists across the world. This book does not provide only current landscape of pigeonpea genomics at international level in terms of tools and strategies employed in genome sequencing, transcriptomics, functional analysis, trait mapping and molecular breeding but also present a road map for accelerating genetics and genomics research for enhancing genetic gains in pigeonpea improvement.

Acknowledgements Authors are thankful to the United States Agency for International Development (USAID); Biotechnology Industry Partnership Programme (BIPP), Department of Biotechnology, Government of India, Science and Engineering Research Board (SERB), Government of India, and ICRISAT for funding various projects related to pigeonpea genomics. This work has been undertaken as part of the CGIAR Research Program on Grain Legumes. ICRISAT is a member of CGIAR Consortium.

References

Pazhamala L, Saxena RK, Singh VK, Sameerkumar CV et al (2015) Genomics-assisted breeding for boosting crop improvement in pigeon pea (*Cajanus cajan*). Front Plant Sci 6:50

Saxena RK, von Wettberg E, Upadhyaya HD, Sanchez V, Songok S, Saxena KB et al (2014) Genetic diversity and demographic history of *Cajanus* spp. illustrated from genome-wide SNPs. PLoS ONE 9:e88568

Varshney RK, Bansal KC, Aggarwal PK, Datta SK, Craufurd PQ (2011) Agricultural biotechnology for crop improvement in a variable climate: hope or hype? Trends Plant Sci 16:363–371

Varshney RK, Chen W, Li Y, Bharti AK, Saxena RK, Schlueter JA et al (2012) Draft genome sequence of pigeonpea (*Cajanus cajan*), an orphan legume crop of resource-poor farmers. Nat Biotechnol 30:83–89

Key Plant and Grain Characteristics and Their Importance in Breeding and Adaptation of Pigeonpea Cultivars

2

K.B. Saxena, Rachit K. Saxena and Rajeev K. Varshney

Abstract

Complexity of yield formation is well understood by all, and there is no specific formula for breeding high-yielding cultivars. Breeders, however for a long time, have relied on selecting various individual traits, which, they thought, will contribute to yield formation directly or indirectly, and in these ways, successes have been achieved in almost all the crops. In pigeonpea, the most important individual plant traits, known to be linked to seed yield are number of pods, primary and secondary branches, and pod-bearing bunches. All these traits are quantitative in nature and have low heritability. In this paper, an attempt has been made to identify various qualitative and quantitative traits related to seed yield, quality, and those preferred for marketing and milling. A brief description about their inheritance and association with yield has also been provided to help breeders in decision-making.

2.1 Introduction

Yield is a product of numerous direct and indirect pathways originating from various traits and their complex interactions among themselves. A number of plant and environmental studies have been conducted in different crops using sophisticated models to understand the process of yield formation. Besides this, various statistical and biometrical methods have been proposed to eliminate/minimize environmental effects to understand the role of specific trait in the determination of yield. In order to select high-yielding genotypes, breeders have attempted to establish relative contribution of individual traits in determining yield. This was done with the help of simple statistical tools such as correlations, regressions, path analyses. However, none of the approaches is foolproof and has their own pros and cons.

K.B. Saxena · R.K. Saxena · R.K. Varshney
International Crops Research Institute for the Semi-Arid Tropics (ICRISAT), Patancheru 502324, India

K.B. Saxena (✉)
Villa 17, NMC Housing, Al Ain, Abu Dhabi, United Arab Emirates
e-mail: kbsaxena1949@gmail.com

R.K. Varshney et al. (eds.), *The Pigeonpea Genome*,
Compendium of Plant Genomes, DOI 10.1007/978-3-319-63797-6_2

Based on number of genes, the traits have been considered a qualitative (1 or 2 genes with major effect) or quantitative (more genes, each with minor effect). In general, the qualitative traits contribute relatively less to yield but are important for resistances, restoration of fertility, and certain market-preferred traits. The heritability of such traits is usually high and has a little or no environmental effect. The quantitative traits, such as plant vigor, pod and branch number, on the other hand, play relatively a greater role in yield formation, but various major or microenvironment often influences their expression. In addition, there are some 'super' traits such as photosensitivity and growth habit (determinate or non-determinate), which are controlled by fewer genes, but when expressed, they alter the phenotype of the plant by altering the expression of other traits. In such events, the estimates of gene action, based on phenotypic data, are biased and useful conclusions cannot be derived. Therefore, as suggested by Byth et al. (1981), the studies related to the estimation of various genetic parameters, and the phenotypic selection should be exercised under the environment for which the cultivar breeding is targeted.

With all these limitations, the trait-based selection is being carried out in almost every crop and it has been effective with different degrees of success. In the present paper, the potential role of different plant and grain characteristics in determining yield and stability of pigeonpea are discussed to assist breeders in selection.

2.2 Productivity Traits

Maturity: Earliness in a crop has always fascinated both researchers and farmers. The traditional pigeonpea cultivars take 6–9 months to mature. Recently, some early lines maturing in 120–130 days were bred, and they succeeded in creating new production niches. Breeders at ICRISAT continued their efforts to reduce its maturity further to help in widening its adaptation. They succeeded in breeding genotypes, which take <50 days to flower, and their maturity is achieved in 80–90 days (Vales et al. 2012). In advanced trials, such genotypes have produced about 800–1000 kg/ha yield with mean per day productivity of 10 kg/ha. Since in pigeonpea earliness is tightly linked to photo-insensitivity (Wallis et al. 1981), such cultivars can help in broadening the adaptation to warm season windows (>20 °C) at higher altitudes (up to 1600 m), wider latitudes (up to 40° N/S), and under short-fallow between two normal crops (Table 2.1).

Table 2.1 Maturity groups of pigeonpea established at 17°N

Broad group	Sub-group	Days to flower[a]	Reference variety	Remarks[a]	Prodn. system
Super early	00	<50	MN 5	Photo-insensitive	Sole crop
Extra early	0	51–60	ICPL 88039	Photo-insensitive	Sole crop
Early	I	61–70	Prabhat	Photo-insensitive	Sole crop
	II	71–80	UPAS 120	Photo-insensitive	Sole crop
	III	81–90	Pusa ageti	Photo-sensitive	Sole crop
	IV	91–100	T. 21	Photo-sensitive	Sole crop
Medium	V	101–110	Maruti	Photo-sensitive	Intercrop
	VI	111–130	Asha	Photo-sensitive	Intercrop
	VII	131–140	ICP 7035	Photo-sensitive	Intercrop
Late	VIII	141–160	Bahar	Photo-sensitive	Intercrop
	IX	>160	MAL 13	Photo-sensitive	Intercrop

[a]Based on field observations on flowering at ICRISAT

Branches, pods, and pod-bearing clusters per plant: In pigeonpea, yield components such as number of branches, pods, and pod-bearing clusters are interlinked and influence the realized yield (Table 2.2). These traits are quantitative in nature and highly influenced by changes in the growing conditions and cropping systems. The heritability estimates of these traits are also low, and selection advance is limited. In spite of their low selection efficiency, pigeonpea breeders have used these traits frequently, but the results in terms of productivity gains are not very encouraging.

Seed size: There is a large variation (2–22 g/100 seeds) for seed size in pigeonpea germplasm. Seed size is an important component of yield in pigeonpea, and its relationship with yield is curvilinear in nature. D. Sharma (Pers. Comm.) studied this relationship in a large number of breeding lines and germplasm at ICRISAT, and he concluded that in the genotypes with seed size of ≤ 10 g/100 seeds, the correlation between yield and seed size was positive. This relationship, however, reversed within the large seeded (≥ 15 g/100 seeds) group of genotypes. Interestingly, there existed no relationship between these two traits in the seed size range of 11–14 g/100 seeds. Traders, millers, and consumers accept this seed size, and most breeders are now working within this range.

Pod size: In pigeonpea, the genetic variation for pod size (=seeds/pod) is large and varies from 2 to 9. In India, the large-podded genotypes are invariably consumed as a fresh vegetable, while the cultivars grown for *dal* purpose generally contain 4–6 seeds/pod. Normally, the traits such as pod size, seed size, pods/bunch, number of secondary branches are negatively associated with number of pods on a plant. The large-podded (8–9 seeds/pod) genotypes have shy pod bearing and suffer with the inherent problem of ovule abortion. In majority of the pods, 1–2 ovules fail to develop into seeds. This problem could be associated with the limited supply of food reserves to the developing ovules. In the small-podded genotypes, there is no issue of ovule abortion.

Plant biomass and harvest index: According to Y. S. Chauhan (Pers. Comm.), the high-yielding both inbred and hybrid cultivars have more or less similar partitioning and harvest indices. He further postulated that high yields recorded in hybrids were primarily due to greater biomass production and relatively with more pod-bearing sites. Bharathi and Saxena (2012) under controlled environment showed that the excessive vigor in the hybrids starts accumulating from very early seedling stage. This means that the plant vigor can be used as selection criterion, but within a given maturity, plant type, cropping system, and availability of moisture during reproductive stage.

In pigeonpea, harvest indices have little or no value for breeders, because their values are low and the estimates of genetic parameters such as genetic variation and heritability within a maturity group are not large enough to carry out any breeding exercise and expect significant genetic advances. In addition, being a perennial plant, its accurate measurement is also very difficult. Besides this, it is also proven that the plant biomass production is highly sensitive to changes in environment and cropping systems. In medium-duration cultivars, it was found to vary between 0.15 and 0.20 (Narayanan and Sheldrake 1979). Natarajan and Willey (1980) also

Table 2.2 Summary of relationships of yield with some key traits

Correlation of yield with	Reported correlations
Days to flower	+ve
Days to mature	+ve, non-sig
Plant height	+ve, −ve, non-sig
Seed size	+ve, −ve, non-sig
Seeds/pod	−ve, non-sig
Branches/plant	+ve
Pods/plant	+ve

reported large difference in the harvest indices; under pure stand of pigeonpea, it was 0.19, as compared with 0.32 in the intercrop. Matters are further complicated by pigeonpea shedding leaves throughout their growth cycle and are not normally included in the measurement.

2.3 Adaptation and Stability Traits

2.3.1 Diseases

Genetic resistances to key biotic and abiotic stresses reduce losses and provide stability to the production.

Fusarium wilt and sterility mosaic are major pigeonpea diseases causing severe yield losses each year. For wilt resistance, both recessive and dominant genes have been reported (Saxena et al. 2012); while for sterility mosaic, the genetics of resistance is unclear. Singh et al. (1983) reported that resistance to sterility mosaic virus was controlled by two dominant and two recessive alleles. Sharma et al. (1984) reported that the resistance to this virus was controlled by four alleles at two major loci. Of these, one each of dominant and recessive alleles together gives immune reaction. In both wilt and sterility mosaic, some prominent biotypes/races have also been observed. For fusarium wilt, the situation with respect to races and their resistance sources is inconclusive. For sterility mosaic disease, three prominent races, specific to area, have been identified. These are designated as Patancheru, Bangalore, and Dholi (Bihar) races, and their resistance sources are well defined.

In addition to wilt and sterility mosaic, Phytophthora stem blight is another potential disease. For this disease also, the existence of races has been established, but their biology and genetics of resistance are unclear. The race situation for Phytophthora blight is still unclear. Alternaria blight, though a minor disease, can cause severe damage in the post-rainy season sowings. A single recessive gene (Sharma et al. 1987) controls the resistance to Alternaria blight.

2.3.2 Insects

Helicoverpa armigera is the most common pod-boring insect of pigeonpea. The annual losses caused by this insect to pigeonpea are estimated to be around US $317 m. The genetic solutions to manage this constraint have not been successful and so far farmers resort to excessive use of chemicals to protect their crops. As an alternate breeding approach, the use of an endotoxin of *Bacillus thuringiensis* (*Bt*) is being tried in pigeonpea at ICRISAT. In pigeonpea, Sharma et al. (2006) reported the development of transgenic for pod borer resistance through direct organogenesis of axillary bud following 72 h co-cultivation with *A. tumefacience*. According to Sharma et al. (2008) although the transgenic pigeonpea plants with *Bt* are available at ICRISAT, the expression of the target genes in the selections for efficiently controlling pod borers under field conditions has been very low, and work is in progress to develop plants with better events.

2.3.3 Waterlogging

Temporary waterlogging in soils with high water-holding capacity poses a serious threat to pigeonpea productivity (Reddy and Virmani 1981). In India alone, about 1.1 m ha of land is waterlogged annually, causing losses of about 25–30% in the productivity (Choudhary et al. 2011). Under waterlogged situations, the useful aerobic bacteria become inactive while their anaerobic counterparts (both facultative/obligate bacteria) become active, and this results in the shortage of oxygen in the soil (Jackson 1990). This adversely affects general plant health. For screening waterlogging tolerance, a reliable technology was developed by Chauhan et al. (2008), and recently, a number of tolerant genotypes have been identified (Sultana et al. 2013). The resistance to waterlogging is controlled by a single dominant gene (Perera et al. 2001).

2.3.4 Drought

Drought is a universal abiotic constraint, and it may affect the crop at early, intermittent, or terminal growth stages with variable intensity (Lopez et al. 1996). In pigeonpea, very little work has been done to understand this constraint, and so far, no genotype with noticeable genetic resistance has been identified.

2.4 Grain Quality Traits

Pigeonpea produces quality grains with 20–22% protein. To produce more protein and meet the requirements, there is a need to breed cultivars with high (20–22%) protein and seed yield as good as traditional cultivars. At ICRISAT, breeding for high protein was taken up using wild species as donor parents. The newly bred lines had protein between 28 and 30% with yield as good as cultivar BDN 1 (Saxena and Sawargaonkar 2015). An estimate of protein yield from this genotype showed (Table 2.3) that the cultivation of such high-protein cultivars, in one hectare additional 100,000 gram protein could be harvested for the farming families.

2.5 Market-preferred Traits

Pigeonpea seeds are non-endospermic with a tightly glued seed coat. To prepare *dal* of good quality with minimum losses, the commercial millers and traders have preferences for various seed traits. These include seed size, shape, and color, besides overall *dal* recovery. Generally, seed size of 10–14 g/100 seeds is preferred for quality *dal* production. Millers easily accept round seeds with white or brown color, but seed lots with mixed colors/size fetch fewer rates in the market. Most millers consider *dal* recovery of about 70% in commercial milling economical. In eastern and southern Africa, the preferred varieties are those with large (>15 g/100 seeds) and white/cream grain color.

2.6 Naked-Eye Polymorphic Markers

Some distinctive morphological traits could be used to ensure genetic purity of breeding lines and cultivars. Such marker traits (Table 2.4) are controlled by recessive genes and popularly called as 'naked-eye polymorphic markers.' Some of such markers identified in pigeonpea germplasm are described herewith.

Obcordate leaf shape: One such important morphological trait is 'obcordate leaf.' This marker is controlled by a single recessive gene (Saxena et al. 2011), and it can be incorporated easily into popular cultivars and hybrid parents. This leaf marker expresses within a month from sowing. The out-crossed hybrid seedlings will have dominant normal (lanceolate) leaves. Such plants can be identified easily with naked eyes for rouging before flowering.

Sesame leaf shape: The plants have long narrow leaves with greenish-yellow color and can be identified easily with naked eyes. Since the sesame leaf trait is controlled by single recessive gene, it can be incorporated easily into the genotypes of interest and can be used as

Table 2.3 Seed and protein yields harvested from high-protein lines

Genotype	Maturity (days)	100-seed wt (g)	Yield (kg/ha)	Protein (%)	Protein yield (g/ha)
HPL 40-5	169	9.6	2100	26.9	452,000 (21.2)
HPL 40-17	169	8.5	2070	26.5	440,000 (18.0)
BDN 1 (C)	168	9.6	2020	23.2	373,000
SEm±	0.9	0.18	160	0.46	–
CV (%)	0.9	3.4	17.3	3.0	–

Source: Saxena and Sawargaonkar (2015); () % advantage over control

Table 2.4 List of potential naked-eye polymorphic traits in pigeonpea

S. no	Plant part	Recessive phenotype	Dominant phenotype
1	Stem	Green	Purple
2		Determinate	Non-determinate
3		Corky	Smooth
4		Single culm	Branching
5		Decumbent	Strait
6	Leaf	Obcordate	Lanceolate
7		Narrow	Lanceolate
8		Sesame	Normal
9	Flower	Cleistogamous	Normal
10		Yellow color	Red color
11	Pod	Green color	Purple color
12	Stature	Dwarf	Tall

marker for maintaining genetic purity in pigeonpea.

Genetic dwarfs: In a well-managed crop, the pigeonpea plants often grow to a height of 2–3 m, and it becomes difficult to manage insects with chemical sprays. The only viable alternative is to tackle this issue by reducing plant height at genetic level. Earlier efforts in this direction succeeded in identifying various dwarfing sources, but the breeders could not succeed in transferring this trait to high-yielding genotypes.

Determinate growth habit: The non-determinate plants have a vegetative terminal bud, which allows the plant to grow in height and spread under adequate moisture conditions. In such plants, the flowers and pods are borne in bunches on the axillary inflorescences arising from nodes. The alternative form of this plant type is designated as 'determinate.' The determinate pigeonpea plants are short in stature and are characterized by reproductive terminal buds. Such plants when they reach flowering stop growing in height. The cultivars with determinate growth habit are not popular because of their greater susceptibility to pod borers. In most cases, the determinate growth habit is controlled by a single recessive gene and can be distinguished easily with naked eyes (Kapoor and Gupta 1991).

Green stem: Genotypes with uniform green stem color were used as 'naked-eye polymorphic marker' in studying the extent of natural out-crossing in pigeonpea (Bhatia et al. 1981). A single recessive gene controls this trait, and its alternate form has dark purple-colored stem.

2.7 Evolution-related Traits

Biologists often ask a question about the evolution of pigeonpea from its wild form to the domesticated types, and some believe that pigeonpea plant is still evolving in nature. The presence of certain plant traits indicates that, in spite of 3000 years of cultivation, the crop is not fully domesticated. The traits such as absence of annual growth cycle, creation of food reserves in stem and other parts, photosensitivity, extensive flower drop, presence of strophiole, and pod shattering in certain germplasm support this view. According to De (1974) and Maesen (1980), the cultivated form of pigeonpea originated from *Cajanus cajanifolius*, a wild relative of pigeonpea, through a single gene mutation. Perhaps a careful comparison of the two species at morphological and genomics levels can through some light on this issue.

Perenniality: Botanically, all the pigeonpea genotypes are perennial. This perennial nature, however, is not very strong across the germplasm, and genetic variation has been observed. In general, the pigeonpea plants survive for

3–5 years. No information is available on the physiology or genetics of this trait.

Photosensitivity: In pigeonpea, long hours of darkness induce flowering, therefore it is classified as a`short-day plant.' Day and night temperatures also interact with prevailing photoperiod to influence the emergence of flowers. The information on the threshold levels for inducing flowering is inconclusive. Wallis et al. (1981) studied the response of extended (16 h) photoperiod on flowering in a range of genotypes and concluded that in pigeonpea, earliness and photo-insensitivity were inversely correlated. Saxena (1981) studied the inheritance of photoperiod reaction in pigeonpea and concluded that three major dominant genes (Ps_1, Ps_2, and Ps_3) were responsible for lateness, and these genes exhibited pleiotropic effects under extended daylength to determine the photoperiod sensitivity. He further concluded that it is not possible to breed late maturing photo-insensitive cultivars in pigeonpea.

2.7.1 Temperature

Inherently, pigeonpea is a warm season pulse and it grows well in the temperature range of 25-35 °C. In both low as well high temperature regimes its growth, flowering and pod set are adversely affected. Under low (<10 °C) temperatures the photosynthesis in the plants is adversely affected due to moisture stress and internal injury causing of cell sap; while the high (>40 °C) temperatures often lead to pollen abortion/sterility and flower drop. In pigeonpea, this has not been an area of serious research and very little and unconfirmed information is available with respect to critical/threshold temperature levels and genetic variation for the tolerance for this abiotic factor.

2.7.2 Need a Section on Temperature Influence

Cleistogamy: Natural out-crossing has been recognized as a major constraint in maintaining genetic purity in pigeonpea. Saxena et al. (1992) selected segregants from an interspecific cross with modified flowers that does not permit natural out-crossing. Since this floral variant is easy to identify and it is controlled by single recessive gene, it offers opportunities to breed cultivars with least or no out-crossing.

Male sterility: In pigeonpea genetic (GMS), cytoplasmic nuclear (CMS), and temperature-sensitive (TGMS) male sterility systems have been discovered. A total of 11 GMS sources were reported from different researchers (see review by Saxena et al. 2010; Saxena 2014). At present, these are not being used in any plant breeding activity, but their maintenance would be a positive step toward conserving biodiversity. Two CMS systems (A_2 and A_4) in pigeonpea have been stabilized, but only A_4 with *C. cajanifolius* cytoplasm is being used in commercial hybrid breeding. The other six CMS sources need to be stabilized for cytoplasmic diversification of pigeonpea hybrids.

Some rare traits: Sometimes, certain rare traits also appear in the crop. This generally happens due to spontaneous recessive mutation and subsequent segregation. Invariably, such traits are lost because of their inability to compete or survive. Most of these traits have no economic value, but can be considered important from academic point of view. Some of such mutants identified at ICRISAT were corky stem (Saxena et al. 1988a, 1988b), open carpel (Saxena et al. 1988a, 1988b), and prostrate or decumbent growth (Saxena et al. 1989) habit. These mutants must be maintained in genetically pure form.

2.8 Traits of Interest in the Wild Species

The cultivated species of crops have originated from their wild ancestors, and it has taken centuries to evolve through natural phenomenon of mutation and selection. Such processes gradually led to species differentiation within a given genera. During this process, some important alleles (mostly recessive and those with minor

Table 2.5 List of important traits available in secondary and tertiary gene pools

Traits lacking in primary gene pool	Potential donor species in secondary gene pool
High protein	*C. scarabaeoides*
	C. albicans
	C. sericeous
Pod borer resistance	*C. scarabaeoides*
Salinity tolerance	*C. sericeous*
CMS inducing cytoplasm	*C. scarabaeoides*
	C. albicans
	C. sericeous
Temperature-sensitive male sterility	*C. sericeous*

effects) were also lost, and the cultivated species lacked these vital traits. Over a period of time, the gene frequencies in highly self-pollinated and those with strong crossability barriers were more or less stabilized, while those with out-crossing continued to maintain such variability.

Based on the crossability barrier, Harlan and de Wit (1971) classified the germplasm into three broad groups and called them gene pools. The primary gene pools consisted of cultivated types and were easy to cross with other sister lines. The secondary and tertiary gene pools were involved all the wild relatives of species. In the former group, the crossable wild species were included, while the non-crossable species constituted the tertiary gene pool. Similar to primary gene pool, in the secondary and tertiary gene pools also, a considerable intra-species genetic variability for different traits exists (Saxena et al. 1990, 1996). This means that for the genetic improvement of cultivated types using its wild relatives, a careful scanning of traits within wild species and their documentation is essential (Table 2.5). From the species representing the secondary gene pool, so far traits such as high protein have successfully been transferred (Saxena and Sawargaonkar 2015). In addition *C. scarabaeoides*, *C. sericeus*, *C. reticulatus*, and *C. cajanifolius* were used to breed cytoplasmic nuclear male sterility systems (Saxena et al. 2010). Due to strong crossability barriers, the transfer of useful traits from the tertiary gene pool is not easy. The only successful example is *C. platycarpus*. Mallikarjuna and Moss (1995) crossed this species with cultivated type using embryo rescue technology. They succeeded in transferring *Phytophthora* blight resistance and earliness to the cultivated types.

2.9 Inheritance of Key Traits

For planning a sustainable genetic enhancement programme, information on gene action of the key traits is essential. In pigeonpea, limited information is available in different maturity groups. The information available from literature (Saxena and Sharma 1990) has been summarized (Table 2.6) for the benefit of the readers. For any detailed information, the original research papers need to be consulted. A perusal of the table shows that both additive as well as non-additive gene actions govern the key traits. For yield enhancement in pigeonpea, both pure line and hybrid breeding programs are in use.

2.10 Genomic Approaches for Trait-Based Breeding

Breeding program in pigeonpea is expected to be enriching through collaborative approaches incorporating genomics interventions. Benefits of combining genomics tools with breeding have been identified in a number of crop species (Varshney et al. 2006). However, in the case of

Table 2.6 Summary information on gene action of some key traits

Trait	Additive	Non-additive	Add. + Non-add.
Days to flower	*		*
Plant height	*	*	*
Plant width	*		*
Days to mature	*	*	*
Pods/plant		*	*
Seeds/pod	*		*
Seed size	*	*	*
Seed yield	*	*	*
Protein %		*	*

Table 2.7 List of traits tentatively identified for genomics research

Trait	Target gene	Donor source	Priority
Wilt resistance	Dominant gene	ICPL 87119	1
SM resistance	Genes from 3 races	ICP 7035	1
Alterneria blight	Recessive gene	ICPL 366	4
Phytophthora blight	Dominant gene	Wild relatives	4
Waterlogging	Dominant gene	ICP 5028/MAL 15	2
High protein	Recessive gene	Wild relatives	3
Fertility restoration	Dominant gene	ICPL 87119	1
Cytoplasmic male sterility	–	ICPA 2039	1
Cleistogamous flower	Recessive gene	ICPL 99050	2
Obcordate leaves	Recessive gene	ICP 5277	2

pigeonpea with few exceptions such as markers for purity testing in hybrids and few preliminary studies on marker trait associations, larger gains from genomics have not realized. Successfully applying genomics in harnessing the genetic gains requires diverse genetic resources, trait phenotyping, genomics tools, bioinformatics, and proof of gene function in crop, i.e., proof of concept. In order to implement genomics in pigeonpea improvement two major milestones have been achieved (1) understanding the desired phenotypic traits in the field (Table 2.7) and (2) developed ample genomics resources including draft genome sequence (Pazhamala et al. 2015). Now, the further challenge is to effectively combine different genomics approaches, integrating information to maximize for pigeonpea improvement. Once marker trait associations established will provide easy/accurate means of selection and transferring desired traits in required genetic backgrounds.

References

Bharathi M, Saxena KB (2012) A comparative study of hybrid and inbred cultivars for germination and other related traits of pigeonpea. J Food Legumes 25:351–354

Bhatia GK, Gupta SC, Green JM, Sharma D (1981) Estimates of natural cross pollination in *Cajanuscajan* (L.) Millsp.: several experimental approaches. In: Proc. Intl. Workshop on Pigeonpeas. International Crops Research Institute for the Semi-arid Tropics, Patancheru, vol 2, pp 129–136

Byth DE, Wallis ES, Saxena KB (1981) Adaptation and breeding strategies for pigeonpea. In: Proc. Intl. Workshop on Pigeonpeas. International Crops Research Institute for the Semi-arid Tropics, Patancheru, vol 1, pp 450–465

Chauhan YS, Silim SN, Kumar Rao JVDK, Johansen C (2008) A pot technique to screen pigeonpea cultivars

for resistance to waterlogging. J Agron Crop Sci 178:179–183
Choudhary AK, Sultana R, Aditya P, Nadarajan N, Jha UC (2011) Breeding for abiotic stresses in pigeonpea. J Food Legumes 24:165–174
De DN (1974) Pigeonpea: evolutionary studies in world crops; diversity, and change in the Indian sub-continent. Cambridge University Press, London, UK. pp 79–87
Harlan JR, de Wet JMJ (1971) Towards a rational classification of cultivated plants. Taxon 20:509–517
Jackson MB (1990) Hormones and developmental changes in plants subjected to submergence and soil water-logging. Aquat Bot 38:49–72
Kapoor RK, Gupta SC (1991) Inheritance of growth habit in pigeonpea. Crop Sci 31:1456–1459
Lopez FB, Johansen C, Chauhan YS (1996) Effects of drought stress on phenology, yield, and yield components timing in short duration pigeonpea. J Agron Crop Sci 177:311–320
Mallikarjuna N, Moss JP (1995) Production of hybrids between *Cajanus platycarpus*, a wild species of pigeonpea. Euphytica 83:43–46
Narayanan A, Sheldrake AR (1979) Pigeonpea (*Cajanus cajan*) as a winter crop in peninsular India. Exp Agric 15:91–95
Natarajan M, Willey RW (1980) Sorghum-pigeonpea intercropping and the effects of plant population density. 1. Growth and yield. J Agric Sci Camb 95:51–58
Pazhamala L, Saxena RK, Singh VK, Sameerkumar CV, Kumar V, Sinha P, Patel K, Obala J, Kaoneka SR, Tongoona P, Shimelis HA (2015) Genomics-assisted breeding for boosting crop improvement in pigeonpea (*Cajanus cajan*). Front Plant Sci 6
Perera AM, Pooni HS, Saxena KB (2001) Components of genetic variation in short-duration pigeonpea crosses under waterlogged conditions. J Genet Breed 55:21–38
Reddy SJ, Virmani SM (1981) Pigeonpea and its climatic environment. In: Proceedings of the International Workshop on Pigeonpeas, vol. 1. ICRISAT Patancheru, India, p 502324
Saxena KB (1981) Report of work carried out at University of Queensland, Australia. pp 44
Saxena KB (2014) Temperature-sensitive male sterility system in pigeonpea. Current Sci 107:277–281
Saxena KB, Sharma D (1990) Pigeonpea Genetics. In: Nene YL, Hall SD, Shiela VK (Eds) The Pigeonpea. CAB Intl. Wallingford, UK, pp 137–158
Saxena KB, Dundas IS, Faris DG, Gupta SC (1988a) Genetics of a corky stem mutant in pigeonpea. J Hered 79:221–222
Saxena KB, Singh AK, Faris DG (1988b) The morphogenesis and inheritance of an open carpel mutant *cd1* in pigeonpea. J Hered 79:75–76
Saxena KB, Gupta SC, Faris DG (1989) Inheritance of prostrate mutant in pigeonpea. Indian J Genet 49:397–398
Saxena KB, Singh L, Reddy MV, Singh U, Lateef SS, Sharma SB, Remanandan P (1990) Intra-species variation in *Atylosia scarabaeoides*(L.) Benth., a wild relative of pigeonpea. Euphytica 49:185–191
Saxena KB, Ariyanayagam RP, Reddy LJ (1992) Genetics of a high selfing trait in pigeonpea. Euphytica 59:125–127
Saxena KB, Rao AN, Singh U, Remanandan (1996) Intra-species variation in *Cajanus platycarpus* for some agronomic traits and crossability. Intl Chickpea Pigeonpea Newslet 3:49–51
Saxena KB, Sultana R, Mallikarjuna N, Saxena RK, Kumar RV, Sawargaonkar SL, Varshney RK (2010) Male sterility systems in pigeonpea and their role in enhancing yield. Plant Breed 129:125–134
Saxena KB, Vales MI, Kumar RV, Sultana R, Srivastava RK (2011) Ensuring genetic purity of pigeonpea hybrids by incorporating a naked-eye polymorphic marker in A and B lines. Crop Sci 51:1564–1570
Saxena KB, Kumar RV, Saxena RK, Sharma M, Srivastava RK, Sultana R, Varshney RK, Vales MI, Pande S (2012) Identification of dominant and recessive genes for resistance to Fusarium wilt in pigeonpea and their implication in hybrid breeding. Euphytica 145:289–294
Saxena KB, Sawargaonkar SL (2015) Genetic enhancement of seed proteins in pigeonpea- methodologies, accomplishments, and opportunities. Intl J Sci Res 4:3–7
Sharma D, Gupta SC, Rai GS, Reddy MV (1984) Inheritance of resistance to sterility mosaic disease in pigeonpea. Indian J Genet 44:84–90
Sharma D, Kannaiyan J, Saxena KB (1987) Sources and inheritance to Alternaria blight in pigeonpea. SABRAO J 19:109–114
Sharma KK, Lavanya M, Anjaiah V (2006) *Agrobacterium*-mediated production of transgenic pigeonpea (*Cajanuscajan* L. Millsp.) expressing the synthetic *Bt cry1Ab* gene. In vitro Cell Develop Biol Plant 42:165–173
Sharma KK, Sreelatha G, Bhatnagar-Mathur P, Saxena KB, Rao DM (2008) Pigeonpea. In: Kole C, Hall TC (Eds) Compendium of transgenic crop plants: tansgenic legume grains and forages. Blackwell Publishing Limited, pp 133–161. ISBN: 978-1-405-16924-0
Singh BV, Pandya BP, Gautam PL, Beniwal SPS, Pande MP (1983) Inheritance of resistance to sterility mosaic virus in pigeonpea. Indian L Genet 43:487–493
Sultana R, Vales MI, Saxena KB, Rathore A, Rao S, Rao SK, Mula M, Kumar RV (2013) Waterlogging tolerance in pigeonpea (*Cajanuscajan.*(L.) Millsp.):

genotypic variability and identification of tolerant genotypes. J Agri Sci 151:659–671
Vales MI, Srivastava RK, Sultana R, Singh S, Singh I, Singh G, Patil SB, Saxena KB (2012) Breeding for earliness in pigeonpea: development of new determinate and non-determinate lines. Crop Sci 52:2507–2516
van der Maesen LJG (1980) India is the native home of pigeonpea. In de Wit HCD (ed) Liber Grantulatorius innonerem. Agriculture University Wageningen, The Netherlands, pp 257–262
Varshney RK, Hoisington DA, Tyagi AK (2006) Advances in cereal genomics and applications in crop breeding. Trends Biotechnol 24:490–499
Wallis ES, Byth DE, Saxena KB (1981) Flowering responses of thirty-seven early maturing lines of pigeonpea. In: Proc. Intl. Workshop on Pigeonpeas. International Crops Research Institute for the Semi-arid Tropics, Patancheru, vol 2, pp 143–150

Botanical Description of Pigeonpea [*Cajanus Cajan* (L.) Millsp.]

3

C.V. Sameer Kumar, S.J. Satheesh Naik, Nidhi Mohan, Rachit K. Saxena and Rajeev K. Varshney

Abstract

Pigeonpea [*Cajanus cajan* (L.) Millspaugh] is an important legume crop of the *papilionaceae* family. It is an often cross-pollinated crop, and breeding principles of both self and cross-pollinated crops are highly effective in its genetic enhancement. Pigeonpea is a hard woody shrub, extensively adaptable to a range of soil types, temperature, and rainfall. It has a deep taproot system extending up to two meters and can grow to a height of four meters. Pigeonpea roots form a symbiotic association with *Brady rhizobium* spp. and perform biological nitrogen fixation. The branching pattern of stem may vary from bush type to compact upright type and is of determinate, semi-determinate, and non-determinate type based on the flowering pattern. The primary leaves are simple, opposite, and caduceus, while the latter ones are pinnately trifoliate with lanceolate to elliptical leaflets. Pigeonpea flowers are zygomorphic, borne on terminal or auxiliary racemes and are normally yellow in color with some variations. It has ten stamens in diadelphous condition with light or dark yellow anthers. The ovary is superior with a long style attached to a thickened, incurved, and swollen stigma. Pigeonpea is an often cross-pollinated crop with an average of 20% cross-pollination. The fruit of pigeonpea is called pod, which is of various colors, with and without deep constrictions. Seeds (with 20–22% proteins and amino acids) can be round or lens shaped, in shades of white and brown color with yellow color cotyledon. Pigeonpea is a widely consumed multi-utility pulse crop, thus the knowledge about the crop botany is vital for modifying it according to future challenges and goals.

C.V. Sameer Kumar (✉) · R.K. Saxena · R.K. Varshney
Genetic Gains, International Crops Research Institute for the Semi-Arid Tropics (ICRISAT), Patancheru 502324 India
e-mail: c.sameerkumar@cgiar.org

S.J. Satheesh Naik
Pigeonpea Breeding, ICAR-IIPR, Kanpur, UP, India

N. Mohan
ANGRAU, Guntur, Andhra Pradesh, India

R.K. Varshney et al. (eds.), *The Pigeonpea Genome*,
Compendium of Plant Genomes, DOI 10.1007/978-3-319-63797-6_3

3.1 Introduction

Cajanus cajan, commonly known as pigeonpea, is a multipurpose drought-tolerant crop cultivated mainly for its edible seeds which are high in dietary protein. It also has household importance and a number of medicinal uses. Apart from human consumption, it is also used as forage, feed, and meal for animals, piggery, and fishery. *Cajanus cajan* is a natural barrier for soil erosion and biological factory for fixing atmospheric nitrogen in soil. Globally, it is grown in more than 80 countries but it is an important grain legume in Asia (India and Myanmar) and Eastern and Southern Africa (Kenya, Tanzania, Malawi, Uganda, and Mozambique). Pigeonpea has a unique place in Indian farming and India accounts for about 68% of the global production. It is the second most important pulse crop next to chickpea, covering an area of 6.66 million hectares in the world, with an average annual production of 4.85 million tons. Its average productivity is 728.5 kg ha^{-1}. The major pigeonpea growing area (5.06 m ha) is in India with a production of 3.29 million tons with the productivity of 649.9 kg ha^{-1} (FAO 2015). The major states in terms of area and production are Maharashtra, Uttar Pradesh, Madhya Pradesh, Karnataka, Gujarat, Andhra Pradesh, Telangana, and Bihar.

Pigeonpea named by various vernacular names viz., Guandul, poroto guandul, porotoparaguayo, sachacafé, falso café, arveja (Argentina); pigeonpea (Australia); guando (Brazil); mu dou (Chinese); pigeonpea, congo pea, red gram (English); pois cajan, poisd'Angole, ambrevade (French); poisd'angole (French-speaking West Africa); straucherbse (German); Puerto Rican bean, pigeonpea (Hawaii); red gram, tur, arhar, dal (India); frijol de árbol (Mexico); Cumandái (Paraguay); ervilha do Congo, feijão, guandu, ervilha de Angola (Portuguese); cachito (spanish); mbaazi (Swahili); duvart (Swedish); pigeonpea, angola pea (United Kingdom); quinchoncho (Venezuela).

3.2 Origin and Geographical Distribution

The name pigeonpea was first used in Barbados where pigeon were fed the seeds of *Cajanus cajan* (Plukenet 1692). Based on the wide genetic variability, Vavilov (1951) reported that India is the center of origin for cultivated pigeonpea and is also been widely cultivated in many African countries, Egypt, and a bunch of Asian countries since prehistoric times. Eastern Africa was considered as center of origin of pigeonpea by several workers owing to its occurrence in wild form (Zeven and Zhukovsky 1975). Based on the occurrence of wild relatives and diversity, van der Maesen (1980) inferred that India is the primary center of origin and Africa is the secondary center of origin for pigeonpea. It is cultivated in wide range of altitude (0–3000 m) (Ripperton and Hosaka 1942; Krause 1921) and the Latitudinal limit is 30° North and South. However, the optimal being 15–20° for most cultivars. Now it has been acclimatized in several tropical and subtropical countries of the world.

3.3 Taxonomical Hierarchy

The genus *Cajanus* belongs to the sub-tribe *Cajaninae*, tribe Phaseoleae, sub-family Papilionoideae, and family Papilionaceae. The genus *Cajanus* has 11 related genera [(1) *Rhynchosia* Lour., (2) *Dunbaria* W., (3) *Dunbaria* A., (4) *Eriosema* D., (5) *Eriosema* C., (6) Reichenb, (7) *Flemingia* Roxb. Ex Aiton., (8) *Paracalyx* Roxb. Ali, (9) *Adenodolichos*, (10) *Baukea,* and (11) *Carissoa* (Mallikarjuna et al. 2011)] and 32 species; (18 species are endemic to Asia, 13 to Australia, and one to West Africa De 1974; Maesen 1980).

It is postulated that the cultivated pigeonpea originated from *Cajanus cajanifolius* by selection for size and vigor of the plant, non-shattering pods, and larger seed size. However, the cultivated *Cajanus cajan* differs from *Cajanus cajanifolius* in floral morphology, pod and seed color, and 100 seed mass (Mallikarjuna et al. 2012).

Based on the genetic cross-compatibility the species of *Cajanus cajan* are distributed into primary gene pool (GP1), which includes the all available germplasm and *C. cajanifolius.* It is freely crossable with the cultivated types and produces fertile hybrids. While the 10 *Cajanus* species that are cross-compatible with *C. cajan* form the secondary gene pool (GP2), the rest of the species, which do not cross with *C. cajan,* are placed in the tertiary gene pool (GP3) (Remanandan 1990) Table 3.1.

3.4 Botanical Descriptors

When pigeonpea seed is sown under optimal moisture and temperature (29 °C–36 °C), the testa of the seed splits open near the micropyle on the second day. The tip of the radical elongates and emerges from the seed coat. Hypocotyl appears as an arch on the third day and continues to grow upward. The hypocotyl turns light purple. The seedling epicotyl elongates three to seven centimeters before the first trifoliate leaf emerges (Reddy 1990). Likewise, the tender seedling grows and gives rise to an erect woody shrub. The plant shows considerable variations in height, ranging from one to four meters with

Table 3.1 Gene pool system of *Cajanus cajan*

Gene pool	Species
GP1	1) *C. cajan* (L.) Millsp., C. cajanifolius
GP2	1) *C. acutifolius* (F. von Muell.) van der Maese 2) *C. albicans* (W. & A.) van der Maesen 3) *C. cajanifolius* (Haines) van der Maesen 4) *C. lanceolatus* (W. V. Fitzg.) van der Maesen 5) *C. latisepalus* (Reynolds and Pedley) van der Maesen 6) *C. lineatus* (W. & A.) van der Maesen 7) *C. sericeus* (Benth. ex Bak.) van der Maesen 8) *C. trinervius* (D.C.) van der Maesen 9) *C. scarabaeoides* (L.) Thouars 10) *C. reticulates* (Dryander) F. von Muell
GP3	1) *C. aromaticus* van der Maesen 2) *C. cinereus* (F. von Muell) F. von Muell 3) *C. crassicaulis* van der Maesen 4) *C. crassus* (Prain ex. King) van der Maesen 5) *C. elongatus* (Benth.) van der Maesen 6) *C. grandiflorus* (Benth. ex Bak.) van der Maesen 7) *C. goensis* Dalz 8) *C. heynei* (W. & A.) van der Maesen 9) *C. kerstingii* Harms 10) *C. lanceolatus* (W. V. Fitzg.) van der Maese, 11) *C. lanuginosus* van der Maesen 12) *C. mareebensis* (Reynolds and Pedley) van der Maesen 13) *C. marmoratus* (R. Br. ex Benth.) F. von Muell 14) *C. mollis* (Benth.) van der Maesen 15) *C. niveus* (Benth.) van der Maesen 16) *C. platycarpus* (Benth.) van der Maesen 17) *C. pubescence* (Ewart & Morrison) van der Maesen 18) *C. rugosus* (W. & A.) van der Maesen 19) *C. villous* (Benth. ex. Bak.) van der Maesen 20) *C. viscidus* van der Maesen 21) *C. volubilis* (Blanco) 22) *C. convertiflorus* F. von Muell.

Fig. 3.1 Full-grown pigeonpea plant

taproots that extend up to two meters into the soil. In most of the types, branching begins from the sixth to the tenth node, *i.e.,* from 15 to 20 centimeters above the ground and covers with full-fledged lush green foliage (Fig. 3.1). The comprehensive botanical description of main parts of pigeonpea plant is given below.

3.4.1 Root System

Under optimal conditions, root growth starts on second day after sowing. Splitting of the testa takes place near the micropyle and radicle emerges and elongation starts from the seed coat. Cambial activity results in secondary thickening (Bisen and 1981). Pigeonpea root system possesses taproot in central with numerous lateral and secondary branches. The taproot becomes thick and woody (Fig. 3.2). The length and spread of the root system is governed by the varietal characters. The erect types produce deeper and penetrating roots where as the spreading ones have shallower and spreading root system (Mahta and Dave 1931). Depending on the varieties, roots may grow deep more than two meters in the soil.

Pigeonpea is nodulated by the cowpea group of rhizobia (*Brady rhizobium* spp.) and forms rhizobia-symbiotic system, mainly on the upper 30 cm of the root system takes active participation. Nodulation starts approximately 15 days after sowing and continues up to 120 days. It declines toward pod filling stage (Kumar Rao 1990). Meristematic zone anchors the development of nodules (Bisen and Sheldrake 1981). The shape of nodules may be oval, elongate, or spherical, and the size varies from 2 to 4 mm. The rhizobia-symbiotic systems play a significant role in improving the fertility and productivity of low nitrogen (N) soils of arid and semi-arid regions and in turn act as a biological factory for N (Sheldrake and Narayanan 1979; Hamdi 1999).

3.4.2 Stem

The nodes are connected by the primary vascular tissues organized into strands, and each strand is associated with a ridge on the stem (Bisen and Sheldrake 1981). Starch accumulation will be noticed in xylem parenchyma and medullary rays during vegetative phase, and during reproductive phase, these will be mobilized for pod development (Sheldrake 1984).

3.4.3 Branches

The branching pattern in pigeonpea depends on genotype, habitat, and spacing of the plants.

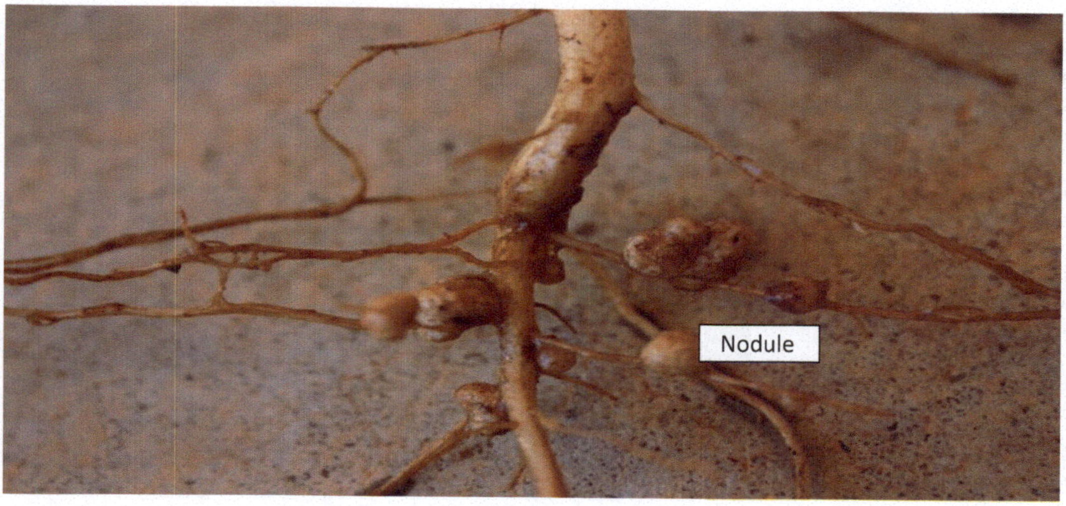

Fig. 3.2 Pigeonpea root system (**a**) nodules on the root branches

Wider spacing may form a bush and at narrow spacing may remain compact and upright. For agronomic purposes, pigeonpea plants can be grouped as compact (erect), semi-spreading (semi-erect), and spreading types. Based on the flowering pattern, it may be determinate or non-determinate (Fig. 3.3b). The determinate type completes the vegetative phase and then enters into the reproductive phase. In this type, the apical bud of the main shoot develops into an inflorescence, and the sequence of inflorescence production is basipetal (developing in the direction of base). The non-determinate type shows continuous vegetative and reproductive phases. In this type, the flowering starts at nodes behind the apex and proceeds both acropetally and basipetally. Another group is semi-determinate between the determinate and non-determinate types. It includes late-maturing genotypes where branching starts from different angles, but most of the pods are at the upper region of the plant.

3.4.4 Leaves

The leaf shape in pigeonpea varied form lanceolate to elliptical in shape. The leaflets are borne on a rachis, which is swollen at the base (pulvinus). The leaf sizes vary from 6 to 17 cm in length and are about the same width. The rachis varies from 2 to 4 cm, and the terminal leaflets are 4–8 cm by 2–3.5 cm. The lateral leaflets are slightly smaller. There is genetic variability in the size, shape, and color of the leaves. The leaves are pubescent with more on the lower than the upper surface. The hair types are simple or glandular. The latter are spherical and contain a yellow oily material, probably responsible for the fragrance of pigeonpea plants (Bisen and Sheldrake 1981).

3.4.5 Inflorescence

In most cultivars, flowers are borne on terminal or auxiliary racemes (4–12 cm) and are carried on a long peduncle (Fig. 3.4a). The raceme inflorescence forms a terminal panicle in non-determinate types and as corymb-shape bunch in the determinate types. These are grouped together at the end of branches in late types and distributed along the branches in early, medium, and indeterminate types (Sharma and Green 1980). The number of racemes plant^{-1} in the pigeonpea world collections ranged from 6 to 915 (Remanandan et al, 1988). Flowering proceeds acropetally (in the direction of apex) both within the raceme and on the branch.

Fig. 3.3 **a** Woody stem of pigeonpea. **b** Different growth habit types in pigeonpea. **c** Trifoliate leaf of pigeonpea

Fig. 3.4 **a** and **b** Flower buds and Full bloom flower of pigeonpea. **c** Dissected flower and Streak pattern of standard petal. **d** Bract of pigeonpea. **e** Calyx

3.4.6 Flowers

The flowers (Fig. 3.4a, b) are clustered at the top of the peduncle. The peduncles are 1–8 cm long. Flowers are normally yellow, however; the main color of the petals could be ivory (green–yellow group 1), light yellow (yellow group 6D), yellow (yellow–orange group 14A), and orange–yellow (orange–red group 31A). Color of streaks on dorsal side of the vexillum (flag) and second color of the wings and keel petals could be red (red group 45 A) or purple (grayed–purple group 186A) (Royal Horticultural Society). The streaks pattern of second color on the dorsal side of the flag (standard petal) ranges from no/sparse streaks to dense/uniform coverage of streaks (Fig. 3.4c).

The bracts (Fig. 3.4d) are small with a thick middle nerve. They are ovate-lanceolate with hairy margins and curved inward to form a boat-like structure to enclose one-to-three young lateral buds. The pedicel is thin, 5–15 mm long, and covered with hair. The flowers are mostly yellow and papilionaceous or completely bisexual and zygomorphic (Sundaraj and Thulasidas 1980). The calyx is gamosepalous with five lobes. The calyx tube is campanulate (bell-shaped) with nerved teeth. The upper two teeth are sub-connate. The lower three are free and spreading (Fig. 3.4e). The upper lobes are paired, free or partly free, and the lower one is the longest.

Corolla: The corolla is zygomorphic (yoke-shaped flowers symmetrical about one plane) and bright yellow. The petals are imbricate and of three prominent types; the standard, wings, and a keel. The standard is broad, large, auricled, and erect (Fig. 3.5a). The wings are obliquely obovate with an incurved claw (Fig. 3.5b). The keel petals are obtuse (round) incurved and boat shaped (Fig. 3.5c). The keel covers the androecium (stamens) and gynoecium (female organs) of the flower.

Androecium: The two halves of the anthers are joined by a relatively large, sterile connective tube that is basi fixed. The anthers are light or dark yellow, dorsi fixed. Of the 10 stamens, four have short filaments and six, including a posterior one, have long filaments. The short anthers have blunt lobes and the long ones pointed lobes. The pollen produced by short stamens is generally used for self-fertilization (Bahadur et al. 1981).

Gynoecium: The ovary is superior, subsessile, flattened dorsoventrally along with style (Fig. 3.5e). It has a very short stalk, densely pubescent, and glandular punctate (dotted or pitted) with two to nine ovules, marginal placentation, monocarpellary, and unilocular. The style is long, filiform, upturned beyond the middle region, and glabrous. It is attached to a thickened, incurved, and capitate (swollen) stigma.

3.5 Pollination

Pigeonpea is an often cross-pollinated crop ranges from 3 to 40%. In a fully developed bud, anthers surround the stigma and dehisce a day before the flower opens. Anthesis in pigeonpea starts from 06.00 h and continues till 16.00 h. The peak anthesis period recorded is between 09.00 and 10.00 h (Sharma and Green 1980). The duration of flower opening also depends on the weather and environment. This varies from 6 to 36 h. Fertilization occurs on the day of pollination.

3.6 Pod/Fruit Development

The fruit of pigeonpea is a pod. During the first week of anthesis, the endosperm undergoes rapid development. The nuclei take up a parietal position, forming a large vacuole in the center of the embryo sac. The embryo sac elongates at the chalazal region and forms a haustorium. The haustorium penetrates into the nucellar tissue. This is instrumental in absorbing food material that is used by the developing embryo. In the cotyledons, synthesis of starch and protein starts about 17 days after pollination and continues for 14 days (Sehgal et al. 1987). In each raceme, 1–5 pods may mature, rarely up to 10. Pods are of various colors (Fig. 3.6); green, purple, dark purple, or mixed green and purple. The seeds per pod range from two to seven, and sometimes up

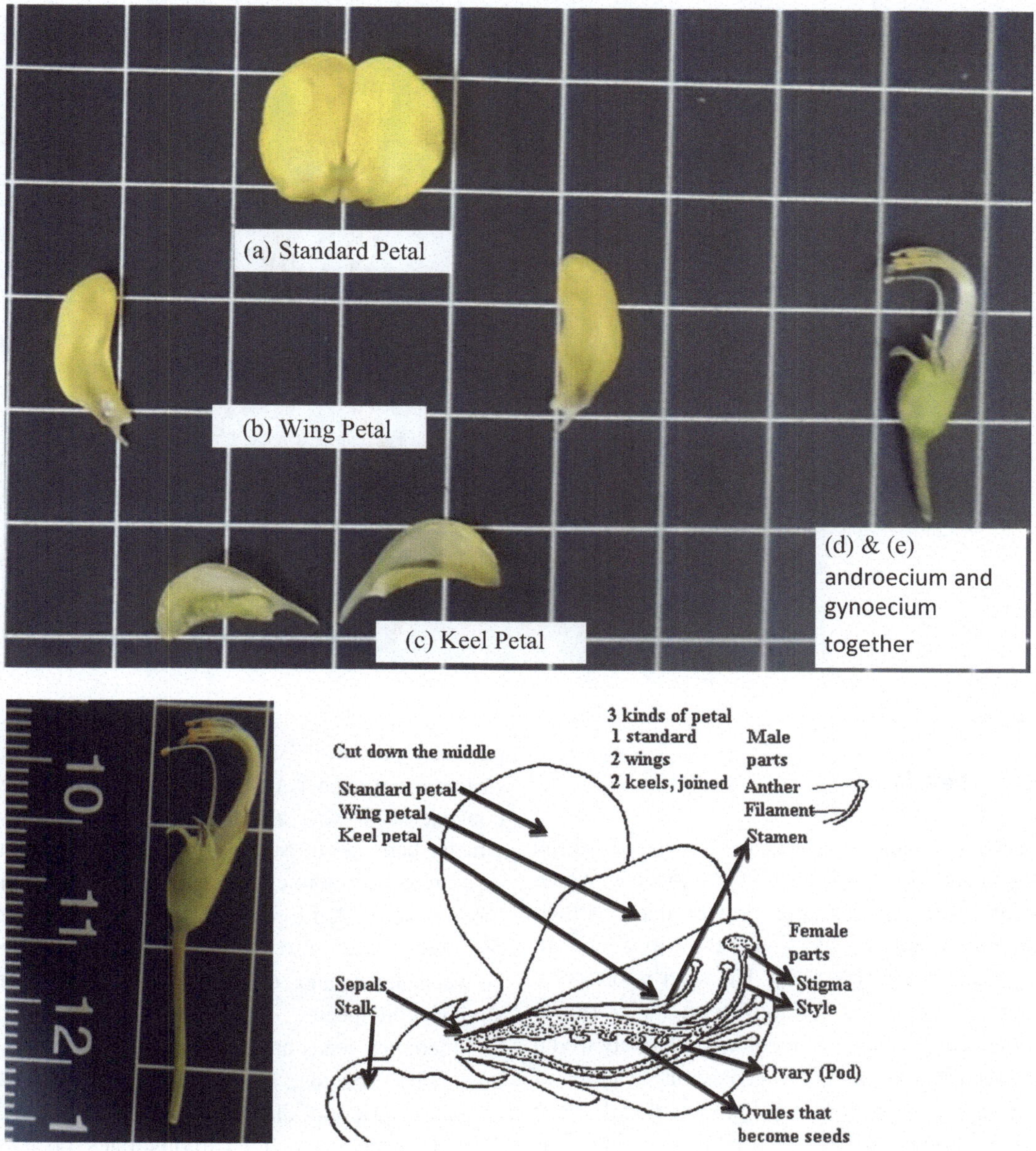

(d) & (e) Close up view of androecium and gynoecium

(f) Schematic detail view of androecium and gynoecium

Fig. 3.5 a Standard Petal. **b** Wing Petal. **c** Keel Petal. **d** and **e** androecium and gynoecium together. **d** and **e** Close up view of androecium and gynoecium. **f** Schematic detail view of androecium and gynoecium

to nine. The seeds are in separate locules, and the cross-walls develop during the first week after fertilization. The pod wall develops more rapidly than the young seeds. Seed development is visible 7 days after pollination. A pod is formed 15–20 days after fertilization. Seeds reach physiological maturity in 30 days and are ready for harvest at lower moisture content in 40 days (Rao and Rao 1974). There is little or no shattering of mature pods in the field.

Fig. 3.6 Pigeonpea pods are of various colors and sizes

3.7 Seeds

Seeds are round or lens shaped, the color of the seed coat varies from dirty white to silver white, light brown to chestnut brown, dark mottled brown and pinkish black, and the cotyledons are yellow colored (Fig. 3.7). The seed is orthodox and can be stored for long term under frozen condition. The seed coat color is relatively trivial character, but it is a consistent feature of pigeonpea evolution that lighter colored testae have consistently been favored in selection. This is frequently the case even when the testa is removed as in the preparation of split peas.

3.8 Chemotaxonomy and Biochemistry

The most productive chemotaxonomic studies carried out to date are those of Ladizinsky and Hamel (1980) and Singh et al. (1981) using electrophoresis. The seed protein separation patterns produced was generally similar in *Cajanus* and former *Atylosia* species, confirming congeneric. It was also observed that species which had been successfully crossed with pigeonpea had more closely similar patterns than those which failed to cross. Variation within the pigeonpea was also found to be less than between itself and *C. cajanifolius*, the most similar wild species.

Pigeonpea seeds are made up of 85% cotyledons, 14% seed coat, and about 1% embryo, and contain a variety of dietary nutrients (Table 3.2). The cotyledons are rich in carbohydrates (66.7%) while a major proportion (about 50%) of seed protein is located in embryo. About one-third of seed coat is made up of fiber. The quantities of important sulfur-containing amino acids such as methionine and cystine range around 1%, and they are present in cotyledons and embryo, while calcium is predominantly present in seed coat and embryo (Saxena et al. 2010). In the review of Toms and Western (1971), its seed is reported as being free of measurable hemagglutinating activity; Liener (1982), in a review of literature

Fig. 3.7 Different colors and sizes of pigeonpea seeds

Table 3.2 Nutrient composition in green seed, mature seed, and dal of pigeonpea

Constituent/cooking time	Green seed	Mature seed	*Dal*
Protein (%)	21.0	18.8	24.6
Protein digestibility (%)	66.8	58.5	60.5
Trypsin inhibitor (units mg^{-1})	2.8	9.9	13.5
Soluble sugars (%)	5.1	3.1	5.2
Flatulence factors (g 100 g^{-1}soluble sugar)	10.3	53.5	–
Crude fiber (%)	8.2	6.6	1.2
Fat (%)	2.3	1.9	1.6
Minerals and trace elements (mg 100 g^{-1})			
Calcium	94.6	120.8	16.3
Magnesium	113.7	122.0	78.9
Copper	1.4	1.3	1.3
Iron	4.6	3.9	2.9
Zinc	2.5	2.3	3.0
Cooking time (min)	13	53	18

(Table re-rewritten from Saxena et al. 2010)

on legume seed toxins and anti-metabolites, mentions the pigeonpea solely on account of a very low cyanide content.

3.9 Ecological Requirement

Cajanus cajan is hard woody shrub, widely adaptable to a range of soil types, temperature and rainfall. Pigeonpea can tolerate the temperatures as high as 35 °C. However, it can be killed by heavy frost. An average annual rainfall between 600 and 1000 mm is most suitable. Perhaps, it can be grown in humid areas, even over 2500 mm of rainfall and is renowned for its drought tolerance. It gives economic yield of seeds in areas where rainfall averages about 400 mm annually. Although it cannot withstand waterlogging, it can be grown in a wide range of soils, as it tolerates low fertility. Some cultivars are tolerant of salinity and aluminum toxicity. A pH range of 4.5–8.4 is tolerated.

3.10 Conclusion

Cajanus cajan is the only cultivated species domesticated from the genus *Cajanus*. It is the unique jewel in rain-fed cropping systems (as sole and intercrop) across the globe. Rich source of protein complements well for a balanced diet with cereals. Versatile use of pigeonpea for human livelihood brought up it as a crop of main land. Pigeonpea is a 'Kalpavriksha' for arid and semi-arid region farmers of the world. Therefore, knowing detailed botany of the crop helps the researchers to modify according to the needs of the future.

References

Bahadur B, Rao MM, Rao KL (1981) Studies on dimorphic stamens and pollen (SEM) and its possible role in pollination biology of *Cajanus cajan* Millsp. Indian J Botany 4:122–129

Bisen SS, Sheldrake AR (1981) The anatomy of the pigeonpea. Research bulletin no. 5. Patancheru, A. P. 502324, India: International Crops Research Institute for the Semi-Arid Tropics, p 24

De DN (1974) Pigeonpea. In: Hutchinson J (ed) Evolutionary studies in world crops: diversity and change in the indian subcontinent. Cambridge University Press, London, UK, pp 79–87

Food and Agriculture Organization (2015) http://faostat.fao.org/database

Hamdi HZ (1999) Rhizobium-legume symbiosis and nitrogen fixation under severe conditions and in an arid climate. Microbiol Mol Biol Rev 63(4):968–989

Krauss FG (1921) The pigeonpea (*Cajanus indicus*): its culture and utilization in Hawaii. University of Hawaii Agri. Expt. Station Bulletin 46, Honolulu, Hawaii, USA

Kumar Rao JVDK (1990) Pigeonpea: nitrogen fixation. In: Nene YL, Hall SD, Sheila VK (Eds) The pigeonpea. CAB International, Wallingford, UK, pp 233–256

Ladizinsky G, Hamel A (1980) Seed protein profiles of pigeonpea (*Cajanus cajan*) and some *Atylosia* species. Euphytica 29:313–317

Liener IE (1982) Toxic constituents in legumes, in chemistry and biochemistry of legumes, Oxford and IBH Publishing Co., New Delhi, pp 217–257

Mahta DN, Dave BB (1931) Studies in *Cajanus indicus*. Memoirs Dept Agric India (Botanic Ser) 19:1–25

Mallikarjuna N, Saxena KB, Lakshmi J, Varshney R, Senthilvel S, Sandhya Rani S, Jadhav DR (2011) Differences between *Cajanus cajan* (L.) Millspaugh and *C. cajanifolius* (Haines) van der Maesen, the progenitor species of pigeonpea. Genet Resour Crop Evol doi:10.1007/s10722-011-9691-8

Mallikarjuna N, Saxena KB, Laxmi J, Varshney R, Srikanth S, Jadhav D (2012) Differences between *Cajanus cajan* and *C. cajanifolius,* the progenitor species of pigeonpea. Genet Resour Crop Evol 59:411–417

Plukenet L (1692) Phytographia 3, Table 213, Figure 3

Raddy LJ (1990) Pigeonpea: morphology. In: Nene YL, Hall SD, Sheila VK (Eds) The pigeonpea. CAB International, Wallingford, UK, pp 47–88

Rao MKV, Rao RG (1974) Gibberellin like substance in developing and germinating seeds of pigeonpea. Indian J Plant Physiol 17(1/2):65–72

Remanandan P (1990) Pigeonpea: genetic resources. In: Nene YL, Hall SD, Sheila VK (Eds), The Pigeonpea. CAB International, Wallingford, UK, pp, 89–115

Remanandan P, Sastry DVSSR, Mengesha MH (1988) ICRISAT pigeonpea germplasm catalogue: evaluation and analysis. Patancheru, A.P. 502324, India: International Crops Research Institute for the Semi-Arid Tropics, p 95

Ripperton JC, Hosaka EY (1942) Vegetation zones of Hawaii. Hawaii Agricultural Experiment Station, University of Hawaii, Honolulu (HI), pp 60 (Bulletin No. 89)

Royal Horticultural Society (Great Britain). Extracts from the proceedings of the Royal Horticultural Society, (http://www.biodiversitylibrary.org/bibliography/79650)

Saxena KB, Kumar RV, Sultana R (2010) Quality nutrition through pigeonpea—a review. Crop Sci 2 (11):1335–1344

Sehgal CB, Gandhi V, Varma B (1987) Histological and histochemical studies on the cotyledons of some legumes. II. Reserve metabolites during seed development. Cytologia 52:847–858

Singh U, Jambunathan R, Gurtu S (1981) Seed protein fractions and amino acids composition of some wild species of Pigeon pea. J Food Sci Technol 18:83–85

Sharma D, Green JM (1980) Pigeonpea. In: Fehr WR, Hadley HH (Eds) Hybridization of crop plants. American Society of Agronomy and Crop Science Society of America, Madison, WI, USA, pp 471–481

Sheldrake AR (1984) Pigeonpea. In: Goldsworth PR, Fisher NM (Eds) The physiology of tropical field crops. John Wiley & Co, Chichester, USA, pp 385–417

Sheldrake AR, Narayanan A (1979) Growth development and nutrient up take in pigeonpea (*Cajanus cajan*). J Agric Sci (UK) 92:513–526

Sundaraj DD, Thulasidas G (1980) Botany of field crops. The Macmillan Company of India Ltd, New Delhi, p 496

Toms GC, Western A (1971) Phytohaemagglutins. In: Harborne JB, Boulter D and Turner BL (Eds) Chemotaxonomy of the Leguminosae. Academic Press, London, pp 367–462

van der Maesen LJG (1980) India is the native home of pigeonpea. In: Arends JC, Boelema G, de Groot CT, Leeuwenberg AJM, Veenman H, Zonen BV (Eds) Libergratulatorius in honorem H.C.D. de Wit landbouwhoge school, Miscellaneous paper no. 19, Wageningen, Netherlands, pp 257–262

Vavilov NI (1951) The origin, variation, immunity and breeding of cultivated plants. Chronica Botanica 13 (1–6):1–366

Zeven AC, Zhukovsky PM (1975) Dictionary of cultivated plants. In: Huxby J (ed) The new systematics. Oxford University, London, UK, pp 549–566

4 Wide Crossing Technology for Pigeonpea Improvement

Nalini Mallikarjuna, Rachit K. Saxena, M. Byre Gowda and Rajeev K. Varshney

Abstract

Pigeonpea (*Cajanus cajan* Millsp,) has ample genetic and genomic information now. It is endowed with rich germplasm in different gene pools. One of the easiest material to use in those are in the primary gene pool, which are closely related to cultivated pigeonpea. It is observed that species placed beyond the primary gene pool are a rich source of genetic variation. They contribute beneficial traits to pigeonpea such as pest or disease resistance, resistance to abiotic stresses, cytoplasmic male sterile systems (CMS) leading to yield improvement, and some novel traits such as homozygous pigeonpea lines. To effectively utilize the immense variation present in the secondary, tertiary, and quaternary gene pool of pigeonpea, a thorough knowledge of crossability and concerted effort is essential.

4.1 Introduction

Genetic variability is the foundation for any breeding program, and breeders look for useful traits first within the primary gene pool and if unsuccessful, then look for the genes available within wild relative of the crop species. These wild species, crossable or non-crossable, may carry genes which may have been lost during natural selection and/or breeding for key economic traits. Mining and utilization of the required trait(s) from wild species is resource intensive, and its success depends on the individual breeder's knowledge, efforts and resources. There are many examples where genes for diseases, insect, quality, etc., have been identified and used in breeding for stability

N. Mallikarjuna · R.K. Saxena · R.K. Varshney
International Crops Research Institute for the Semi-Arid Tropics, Patancheru 502324, India

N. Mallikarjuna (✉)
#213, Esha Krupa, 9th Cross 3rd Stage, Gokulum, Mysore, Karnataka, India
e-mail: mysorenalinicgiar@yahoo.com

M. Byre Gowda
All India Co-ordinated Research Project on Pigeonpea, University of Agricultural Sciences, GKVK, Bengaluru, Karnataka, India

R.K. Varshney et al. (eds.), *The Pigeonpea Genome*, Compendium of Plant Genomes, DOI 10.1007/978-3-319-63797-6_4

and developing new genetic system such as male sterility. The development of technologies such as transformation, somatic hybridization, tissue/embryo culture has made it possible to utilize genes from the wild species separated by vast hybridization barriers.

In recent time, the change in agricultural environment is making subsistence agriculture more difficult and unpredictable. To overcome such threats as drought, high temperature, and emergence of new biotypes of insects and diseases necessitates the search of new genes from the germplasm. The wild species have greater probability of harboring the genes which can sustain the climate changes because over the time they have survived under diverse environments and may carry special survival mechanisms, not available among the cultivated types. Pigeonpea belongs to the subtribe *Cajaninae* which contains 13 genera. The earlier taxonomists considered genus *Atylosia* and *Cajanus* different but closely related. Subsequently, van der Maesen (1980) merged the two genera together and identified as genus *Cajanus*. At present therefore, this genus contains 32 species. Of these, 18 are endemic to Asia, 13 to Australia, and one to Western Africa. The other related genera are *Rhynchosia*, *Dunbaria*, *Flemingia*, *Paracalyx*, *Eriosema*, *Adenodolichos*, *Bolusafra*, *Carissoa*, *Chrysoscias*, and *Baukea* (van der Maesen 1986).

Pigeonpea originated about 3500 years ago, and its landraces contain a lot of variability (Remanadan 1990), but the recent genomics studies showed lack of genetic diversity within primary gene pool, and this is a matter of concern to breeders engaged in the genetic improvement of the crop. Hence, the viable option appears to be the utilization of its wild relatives from secondary, tertiary, and quaternary gene pools using appropriate gene transfer techniques.

4.2 Primary Gene Pool

Considerable progress has been made in pigeonpea improvement by using variability within the cultivated species. In spite of the large germplasm collection in the primary gene pool, it is not widely used (Wright 1997) as information on the presence of useful traits is not easily available and necessitating an extended period of research whenever utilized (Goodman 1990). To overcome these issues, core and mini core collections have been developed (Upadhyaya et al. 2006). Variation within the primary gene pool is of importance as they are easy to use with quicker gains and can be directly released as cultivars. Progress has been made in the utilization of material from the primary gene pool. Varieties BDN-1 and Maruthi are selections from pure-line breeding which are popular even today (Bantilan and Joshi 1996). Development of high yielding varieties such as ICPL 87, ICPL 151, Prabhat, T 21, Pusa Ageti, CO 5, and JA 3 has also been reported (Singh et al. 2005). In pigeonpea, 57 ancestors were used to develop 47 varieties. The top 10 ancestors contributed 48% to the genetic base of the released varieties (Kumar et al. 2003), thus narrowing the genetic base of the crop.

4.3 Secondary Gene Pool

Compatible wild relatives of pigeonpea which are placed in the secondary gene pool do not need specialized techniques in the crossability experiments in majority of the cases, with a few exceptions (Mallikarjuna et al. 2011a).

4.3.1 Cytoplasmic Male Sterile Systems

Cytoplasmic male sterile systems were developed for pigeonpea exploiting the cross-pollination mechanism and utilizing wild *Cajanus* species. So far, nine CMS systems have been reported utilizing wild relatives of pigeonpea (Mallikarjuna et al. 2012; Srikanth et al. 2015). Of these, seven have been developed utilizing wild relatives from secondary gene pool as the female parent. Two systems have cultivated pigeonpea cytoplasm with the utilization of wild species as the male parent (Mallikarjuna and Saxena 2005; Srikanth et al. 2015).

4.3.2 Cleistogamy

Insect-aided natural outcrossing in pigeonpea is a universal event (Saxena et al. 2016), and it leads to rapid contamination of pure lines. The breeders were on the lookout for a genetic trait that could protect the genetic purity at no cost basis. Saxena et al. (1992) discovered a floral modification that inhibits cross-pollination under open fields. This trait, selected from the cross *C. cajan* × *C. lineatus* and identified as "partially cleistogamous" flower, restricts natural outcrossing to less than 1%. This trait is controlled by a single recessive gene and thus easy to handle in pedigree breeding programs. Its stability across the environments makes it an ideal trait for incorporation in future cultivars.

4.3.3 High Protein

High protein breeding lines were developed from *C. sericeus*, *C. albicans,* and *C. scarabaeoides*. Significant positive correlation between seed size and protein content was observed in lines derived from *C. scarabaeoides*. Lines HPL 2, HPL 7, HPL 40, and HPL 51 are some of the high protein and high seed weight lines derived from wild species (Saxena et al. 1987). More recently, crosses between pigeonpea and *C. acutifolius* yielded progeny with high seed weight. High seed weight accompanied by beige seed color is a desirable trait (Jadhav et al. 2012).

4.3.4 Insect Resistance

C. acutifolius, a wild relative from secondary gene pool and native of Australia, can be crossed with pigeonpea as a one-way cross. The reciprocal cross using *C. acutifolius* as the female parent aborts to give rise to immature seeds. In vitro, interventions are necessary to obtain hybrid plants (Mallikarjuna and Saxena 2002). Advanced generation population from the cross utilizing *C. acutifolius* as the pollen parent has shown resistance for pod borer damage (Mallikarjuna et al. 1997; Jadhav et al. 2012), variation for seed color, and high seed weight. Some of the lines showed high level of resistance to pod borers, pod fly, and bruchid under unprotected field conditions (Jadhav et al. 2012). Bruchid resistance (Jadhav et al. 2012) is an important trait for pigeonpea seeds under storage as resistance to the pest has not been observed in cultivated pigeonpea.

Another species from secondary gene pool, namely *C. lanceolatus,* was crossed successfully with cultivated pigeonpea at ICRISAT and progeny lines developed (Srikanth et al. 2013). F_1 hybrids flowered but some of the hybrids were pollen sterile, and in the rest of the hybrids, pollen fertility varied from 25 to 55%. All the hybrids were female fertile. Progeny lines developed from the cross were screened for bruchid resistance. Bruchid growth and survival was inhibited in the lines derived from *C. lanceolatus*. Some of the lines showed delayed bruchid growth and delayed life cycle thus showing antibiosis mechanism of resistance to bruchids. Lines were screened for protein content, and some of the lines showed higher protein content than both their parents. Further, biochemical analysis showed higher content of proteinase inhibitor activity in some of the lines (Srikanth et al. 2017). Previously, Satishkumar (1985) had attempted crossing pigeonpea with *C. lanceolatus* but obtained sterile hybrids which did not flower and remained in the vegetative stage.

4.3.5 Water Logging

Some of the advance generation lines derived from *C. acutifolius* were screened for water logging by germinating them and later growing them under water logged conditions. A few lines grew under water logged conditions, and formation of lenticels was observed in the region above the water surface. The region gave rise to roots which entered the soil through the water surface. This shows that some of these lines may survive water logged conditions as seen in some pigeonpea lines (Hingane et al. 2015).

4.4 Tertiary Gene Pool

There are 20 wild species in the tertiary gene pool of pigeonpea (Mallikarjuna et al. 2011a). Until now, two wild *Cajanus* species from this gene pool have been successfully crossed and traits of interest transferred (Mallikarjuna et al. 2011b). *Cajanus platycarpus* was successfully crossed utilizing hormone-aided pollinations and in vitro interventions (Mallikarjuna et al. 2011a) to obtain hybrids. Progeny lines showed variation for days to flower, growth habit, seed weight and number, seed color, resistance to pod borer, pod fly, bruchids, fusarium, and sterility mosaic disease and CMS (Mallikarjuna et al. 2011b, 2012). Some chasmogamous lines (Cherian et al., 2006) were identified in CMS lines, a trait favoring total cross-pollination. Hence, utilizing *C. platycarpus* not only broadened the genetic base of pigeonpea but it was possible to introgress useful traits.

More recently, another species from tertiary gene pool, namely *C. volubilis,* was crossed with pigeonpea (Mallikarjuna et al. 2014). In F_2 generation, extra short duration lines were recovered in 2011. These lines flowered earlier than the short duration cultivar ICPL 85010 which was the female parent of the cross. Dwarf growth habit coupled with determinate and semi-determinate plants was observed. In the determinate types, the number of pods per inflorescence and the number of inflorescence was more than that observed in the extra-early and determinate cultivar MN5. Allele-specific marker assay developed for the SNP (T/A) was used to genotype 21 F_2 progeny derived from *C. volubilis*. Two bands (one common band (848 bp) and one allele-specific band (734 bp/167 bp) appeared in all the samples. The degenerated common primers (TFL1_PCR_CF and TFL1_PCR_CR) amplified 848 bp-specific fragment, specific for determinate types, among all genotypes (Mir et al. 2014). As short duration is a desirable trait in pigeonpea improvement, and there is an emphasis to look for this trait while breeding for better pigeonpea especially for higher altitudes and latitudes. Although early flowering with short stature is available in cultivated pigeonpea germplasm, the source identified in *C. volubilis* derivatives may have a different genetic background, as *C. volubilis* is a tertiary gene pool species.

4.5 Quaternary Gene Pool

There are 11 related genera, namely *Rhynchosia, Flemingia, Dunbaria, Erisema, Paracalyx, Adenodolichos, Bolusafra, Carissoa, Chrysoscias,* and *Baukea* including *Cajanus* under the subtribe Cajaninae. Many of these genera are classified as underexploited legumes. *Rhincosia* is one such example as it harbors important nutritional and therapeutic properties (Drabu et al. 2011), with the presence of phytochemicals such as alkaloids, glycosides, anthraquinones, carotenoids, coumarins, dihydrochalcones, fatty acids, flavonoids, steroids and triterpenoids (Bakshu and Venkataraju 2001). Some species of *Rhynchosia* are used in human and animal diet (Oke et al. 1995). Many of the tribal communities in India soak the seeds in water and consume the seeds after boiling and decanting many times to get rid of unwanted constituents (Murthy and Emmanuel 2011). Many of the *Rhynchosia* species are known to exhibit antitumor and thus curative properties. Normally, during cancer treatment iron deficiency and anemia are major issues. It was observed that treatment with *Rhynchosia* seeds restored hemoglobin (Hb) count, RBC and WBC count to normal levels. These traits are also important in the treatment of dengue fever, an ailment caused by a variety of mosquito bite. With the interest in dietary flavonoids, suppression of cancer, and treatment in dengue fever, progeny lines developed from pigeonpea and *Rhynchosia* cross would be an asset in pigeonpea improvement.

None of the genera in the quaternary gene pool have been successfully crossed with pigeonpea until now. Among the genera in the quaternary gene pool, *Rhynchosia* was selected to initiate crossing/introgression/gene transfer experiments as it had many desirable properties as listed above. It was possible to successfully cross *Rhynchosia* with pigeonpea through hormone-aided pollinations (Mallikarjuna et al. 2014). The success rate of crossing *Rhynchosia* was low not exceeding 1–2%, but it was possible to obtain hybrids. Screening the hybrids with molecular markers confirmed the hybridity (Mallikarjuna et al. 2014). Although the initial process of crossing *Rhynchosia* with pigeonpea was challenging, nevertheless, hybrids were obtained. They were fertile, and it was possible to obtain self and backcross progenies. Experiments to screen and study the progeny lines for different traits/constraints are in progress.

4.6 Diversity in Expression of Traits Due to Alien Introgression

4.6.1 Segregation of Traits Following Mendelian Pattern of Inheritance

Wild relatives from secondary gene pool cross with cultigen following the Mendelian pattern of inheritance. This meant the F_1 produced, broadly showed the expression of both the parental species in the ratio of 1:1. When the F_1 was backcrossed the recurrent parent, the progeny lines segregated to traits in the ratio of 3:1, again following the Mendelian pattern of segregation. Similar results were observed when a tertiary gene pool species *C. platycarpus* was crossed with the cultigen (Mallikarjuna et al. 2011b) DArT analysis of the BC4F1 lines, where the recurrent parent was the cultigen, showed that it followed the Mendelian pattern of inheritance in having approx. 94% *C. cajan* genome/DNA, and the remaining was that of *C. platycarpus* (Mallikarjuna et al. 2011a).

4.6.2 Segregation Pattern Showing Genomic Dominance

Cajanus volubilis, a wild relative in the tertiary gene pool of pigeonpea, produced F_1 hybrids when crossed with the cultigen (Mallikarjuna et al. 2014). The F_1 hybrid showed traits of both the parental species, i.e., of both *C. volubilis* and the cultigen, and the hybridity was further confirmed by molecular SSR analysis (Mallikarjuna et al. 2014). F_1 hybrid was selfed to produce F_2 progeny lines. All the plants from F_2 onward showed genomic disequilibrium by resembling the cultigen with respect to morphological traits. None of the plants had any morphological traits or characters of *C. volubilis*. Molecular analysis of F_2 plants confirmed that they resembled the cultigen. Such a phenomenon has not been observed in pigeonpea-wide crosses using compatible wild relatives from secondary gene pool (Mallikarjuna and Saxena 2002). Genomic asymmetry was observed in the present cross as distantly related genomes of *Cajanus,* i.e., of the cultigen and that of *C. volubilis,* were brought together for the first time through hybridization. Bringing together distantly related genomes involves radical and rapid mode of speciation by means of interspecific hybridization. It has been observed in wheat when different genomes were brought together there was predominance of one genome over the other (Flagel et al. 2009; Rapp et al. 2009). In wheat A and B genome hybrids, B genome exhibited more or higher marker polymorphism than the A genome (Chao et al. 1989). Genome-wide transcriptome analysis in synthetic *Arabidopsis* allotetraploids showed that expression patterns from one genome could be dominant over the other (Wang et al. 2006a, 2006b). Pumphery et al. (2009) observed that some of the genes were similar to one of the parental genomes in synthetic hexaploid wheat. In *C. volubilis* progeny lines too, predominant expression of one of the parental genome over the other was observed and this can be explained taking examples of wheat, *Arabidopsis,* etc., that when distantly related genomes are brought together, there is silencing of the expression of the other genome completely or partially, although the DNA from the other parent is present, at least for many of the morphological traits, female parental genome expression is obvious.

Genome asymmetry has been observed not only for the expression of morphological traits but for other important traits, in a few crops when different genomes are brought together by hybridization. Genome asymmetry in the control of storage proteins has been observed in wheat (levy et al. 1988). In *Gossypium hirsutum*, genome asymmetry was found in the accumulation of seed storage proteins (Hu et al. 2011). Genome asymmetry in the control of agronomic, disease, and pest resistance traits has also been observed in wheat (Feldman et al. 2012).

There were unsuccessful attempts in the past to cross pigeonpea with *C. volubilis* pollen (Pundir and Singh 1985). This is the first report of successful interspecific hybridization between pigeonpea and *C. volubilis*. Morphological traits of the F_1 hybrid were more skewed toward the female parent, and such phenomenon is not new when distant genomes are made to come together through wide hybridization. Molecular analysis confirmed its hybridity. *C. volubilis* is a wild relative of pigeonpea placed in its tertiary gene pool (Mallikarjuna et al. 2011b; Bohra et al. 2010). Genomic studies have also shown its distant relationship with cultivated pigeonpea (Pangaluri et al. 2007). Genome-wide transcription analysis in synthetic *Arabidopsis* allotetraploids showed that expression patterns from one genome could be dominant over the other genome (Wang et al. 2006a, 2006b). Pumphery et al. (2009) found that a small percentage of hybrids between wheat and synthetic hexaploids was similar to one of the parents. We report for the first time in pigeonpea that such a phenomenon is taking place in the hybrid between *C. cajan* (cultivated pigeonpea) and *C. volubilis* with the morphology of the F_2 hybrids skewed toward the female parent. Genetic control in storage proteins has been observed in allopolyploid wheat. Galili and Feldman (1984) showed that inactivation of endosperm protein is brought about by an inter-genomic suppression. Wheat genome-driven control of some agronomic, pest, and disease resistance was observed in wheat. Peng et al. (2000) observed that R-gene cluster in the B genome of wheat and high marker clustering in the B genome than the A genome is the result of expression of genome asymmetry. The explanation for ability of one genome to suppress the activity of genes in another in newly formed hybrids with different genomes may be to prevent defective organ formation/phenotype. This may be a protective mechanism to obtain viable plants.

4.7 Development of Homozygous Lines

4.7.1 Tissue Culture

Haploids are plants that contain gametic number (n) of chromosomes. Haploids and consequent production of “double haploids” (DH), obtained by doubling of chromosomes either spontaneously or induced via chemical methods, have since been applied to many crop improvement programs, and protocols are available for more than 200 plant species (Bhojwani and Razdan 1996). DH is the fastest route to obtaining homozygosity and can be developed in a single laboratory-based generation. They are equivalent to inbred lines developed by conventional breeding program requiring 6–7 generations of selfing to achieve satisfactory level of homozygosity (Mallikarjuna et al. 2005). Owing to their homozygous and true breeding nature, they are of importance in plant breeding. But legumes are considered to be recalcitrant to DH production with a few successful examples such as *Medicago truncatula* (Lanas et al. 2006), *Glycine max* (Moraes et al. 2004), *Pisum sativum* (Ochatt et al. 2009), and *Lupinus angustifolius* (Kozak et al. 2012).

Except for one report (Kaur and Bhalla 1998), haploid/doubled haploids have not been reported in pigeonpea in spite of multiple and concerted efforts later on. Homozygous or pure lines have gene sets which are exactly identical or two identical haploid genomes. Homozygous lines or doubled haploid plants once developed will be of great significance in pigeonpea, which is a partially cross-pollinated crop. Cross-pollination induces variability which is a desirable trait in pigeonpea, but not always. In certain breeding experiments such as CMS (cytoplasmic male sterility), variability or heterozygosity is not desirable in the development of restorers and to exploit heterosis. Homozygosity can be achieved by anther culture, but this technology does not work for pigeonpea in spite of concerted and repeated efforts (Croser et al. 2006).

4.7.2 Wide Crosses

Utilizing related genera *Rhincosia*, it has been possible to obtain hybrids between *Rhyncosia* species and cultivated pigeonpea (Mallikarjuna et al. 2014). Although the F_1 was a true hybrid, subsequent generation progenies resembled pigeonpea. Molecular analysis of F_2 progeny lines revealed complete homozygosity for all the loci tested. Hence, the resultant plants were completely homozygous. In 2012 field evaluation, F_4 progeny lines showed complete resemblance to the female parent, i.e., to cultivated pigeonpea. Many distantly related species are difficult to cross and when forced to cross, get their genome eliminated. Although this has not been reported for any legume species, in cereal crops such as barley, homozygous plants are obtained in one step by crossing barley with *Hordeum bulbosum*. The method called the bulbosum technique, first reported by Kasha and Kao (1970), is now a routine technique of haploid production in barley through chromosome elimination by wide cross technology. The chromosome elimination phenomenon is quite prevalent among wide crosses between wheat and *H. bulbosum* as well (Barclay 1975). An added advantage in such cases is that there is spontaneous doubling of female parental chromosomes as observed in barley and in pigeonpea too. Mallikarjuna et al. (2005) reported multicellular microspores in most of the anther in a chickpea wide cross *C. arietinum* × *C. pinnatifidum*. If such anthers/microspores are cultured, it may be possible to obtain haploid plants in chickpea. This could be yet another method of haploid production through wide cross technology.

Homozygous plants offer a promising alternative to recurrent selfing for many years/generations, for rapid inbred line development. Even if the inbred lines are developed after repeated selfing, it is known that certain regions of the genome retain heterozygosity (Nair et al. 1995). In none of the leguminous crops, this phenomenon of homozygosity through wide crosses has been reported.

Normally, a diploid will have a combination of different haploid genome/s obtained from each parent through gametic fusion in sexual reproduction. When homozygous lines are developed through wide hybridization, there is preferential elimination of the paternal genome and doubling of the maternal genome chromosomes, thus creating 100% homozygosity in one step. Dominance is eliminated in homozygosity, and much less progeny is needed in the study. For example, for two desired genes, using homozygous lines, one needs to grow four homozygous lines instead of growing 16 when selfed to obtain desired genotype. For four desired genes, using homozygous route, one would need to grow 16 homozygous lines, instead of growing 256 through the selfing route.

4.8 Conclusion

No other food legume crop has been investigated for alien introgression and succeeded in crossing wild relatives from all the gene pools, namely secondary, tertiary, and quaternary gene pools. Pigeonpea is one crop where tremendous progress has been made to cross wild *Cajanus* species from different gene pools and introgress genes/traits successfully. With these successes, it can no more have a narrow genetic base. With the advances in pigeonpea genomics, and a major effort in sequencing the crop, and success in wide crosses in pigeonpea, it has emerged from being labeled as an orphan crop to a trend setter. Recent successes in wide crosses show that it is possible to introduce desirable traits such as pod borer resistance, develop CMS systems, develop lines with multiple disease and pest resistance, change plant type, and increase seed weight and yield. An added advantage is the availability of homozygous lines for any applied research.

References

Bakshu L, Venkataraju RR (2001) Antimicrobial activity of Rhincosia beddomei. Fitoterpia 72:579–582

Bantilan MCS, Joshi PK (1996) Returns to research and diffusion investments on wilt resistance in pigeonpea. Impact series (No. 1) Patancheru 502 324, Andhra Pradesh, India. International Crops Research Institute for the Semi-Arid Tropics, p 36

Barclay IR (1975) High frequencies of haploid production in wheat (Triticum aestivum) by chromosome elimination. Nature 256(5516):410–411

Bhojwani SS, Razdan M (1996) Plant tissue culture: theory and practice, a revised edition, Stud Plant Sci 5:167–214

Bohra A, Mallikarjuna N, Saxena K, Upadhyaya HD, Vales I, Varshney RK (2010) Harnessing the potential of crop wild relatives through genomics tools for pigeonpea improvement. J Appl Biol 37:83–98

Chao S, Sharp PJ, Worland AJ, Koebner RMD, Gale MD (1989) RFLP-based genetic maps of wheat homoeologous group 7 chromosomes. Theor Appl Genet 78:495–504

Cherian CA, Mallikarjuna N, Jadhav DR, Saxena KB (2006) Open flower segregants selected from Cajanus platycarpus crosses. J SAT Agric Res 2:1–2. ISSN 0973-3094

Croser JS, Lulsdorf MM, Davies PA, Clarke HJ, Bayliss KL, Mallikarjuna N, Siddique KHM (2006) Towards doubled haploid production in Fabaceae: progress, constraints, and opportunities. Critical Rev Plant Sci 25:139–157

Drabu S, Chaturvedi S, Sharma M (2011) Analgesic activity of methanolic extract from aerial parts of Rhynchosia capitata DC. Int J Pharm Technol 3:3590–3600

Feldman M, Levy AA, Fahima T, Korol A (2012) Genomic asymmetry in allopolyploid plants: wheat as a model. J Exp Bot 63:5045–5059. doi:10.1093/jxb/ers192

Flagel LE, Chen L, Chaudhary B, Wendel JF (2009) Coordinated and fi ne scale control of homoeologous gene expression in allotetraploid cotton. J Hered 100:487–490

Galili G, Feldman M (1984) Intergenomic suppression of endosperm protein genes in common wheat. Can J Genet Cytol 26:651–656

Goodman MM (1990) Genetic and germplasm stocks worth conserving. J Hered 81:11–16

Hingane AJ, Saxena KB, Patil SB, Sultana R, Srikanth S, Mallikarjuna N, Vijaykumar R, Sameer Kumar CV (2015) Mechanism of water-logging tolerance in pigeonpea. Indian J Genet Pl Br (The) 75(2):208

Hu G, Housron NL, Pathak D, Schmidt L, Wendel JF (2011) Genomically biased accumulation of seed storage proteins in allopolyploid cotton. Genetics 189:1103–1115

Jadhav DR, Mallikarjuna N, Sharma HC, Saxena KB (2012) Introgression of Helicoverpa armigera resistance from Cajanus acutifolius—a wild relative from secondary gene pool of pigeonpea (Cajanus cajan). Asian J Agric Sci 4(4):242–248

Kasha KJ, Kao KN (1970) High Frequency haploid production in Barley (Hordeum vulgare L.). Nature 225(5235):874–876

Kaur P, Bhalla JK (1998) Regeneration of haploid plants from microspore culture of pigeonpea (Cajanus cajan L.). Indian J Exp Biol 36:736–738

Kozak K, Galek R, Waheed MT, Sawicka-Sienkiewicz E (2012) Anther culture of Lupinus angustifolius: callus formation and the development of multicellular and embryo-like structures. Plant Growth Regul 66 (2):145–153

Kumar S, Gupta S, Chandra S, Singh BB (2003) How wide is the genetic base of pulse crops? In: Ali M, Singh BB, Kumar S, Dhar V (Eds) Pulses in new perspectives. Proceedings of the national symposium on crop diversifycation and natural resource management, Indian Institute of Pulses Research, Kanpur, UP, India, pp 211–221

Lanas I, Gallego P, Martin L, Fernandez J, Alonso A, Rosello JE, Blazquez A, Villalobos N, Guerra H (2006) In vitro Culture of Medicago arborea L. Anthers: initial response. Plant Growth Regul 49 (1):49–60

Levy AA, Galili G, Feldman M (1988) Polymorphism and genetic control of high molecular weight glutenin subunit in wild tetraploid wheat Triticum turgidum var. dicoccoides. Heredity 61:63–72

Mallikarjuna N, Saxena KB (2002) Production of hybrids between Cajanus acutifolius and C. cajan. Euphytica 124(1):107–110

Mallikarjuna N, Saxena KB (2005) A new cytoplasmic male-sterility system derived from cultivated pigeonpea cytoplasm. Euphytica 142(1–2):143–148

Mallikarjuna N, Sharma HC, Upadhyaya HD (1997) Exploitation of wild relatives of pigeonpea and chickpea for resistance to Helicoverpa armigera. SAT eJournal 3(1):4–7

Mallikarjuna N, Jadhav DR, Clarke H, Coyne C, Meulhbauer FJ (2005) Induction of androgenesis as a consequence of wide crossing in chickpea. J SAT Agric Res 1(1):1

Mallikarjuna N, Saxena KB, Jadhav DR (2011a). Cajanus. In: Kole C (ed) Wild crop relatives: genomic and breeding resources, Springer, Berlin/Heidelberg, vol 10, pp 21–33

Mallikarjuna N, Senthivel S, Jadhav DR, Saxena K, Sharma HC, Upadhyaya HD, Rathore A, Varshney R (2011b) Progress in the utilization of Cajanus platycarpus (Benth.) Maesen in pigeonpea Improvement. Plant Breeding 130(5):507–514

Mallikarjuna N, Jadhav DR, Saxena KB, Srivastava RK (2012) Cytoplasmic male sterile systems in pigeonpea with special reference to A7 CMS. Elect J Plant Breed 3(4):983–986

Mallikarjuna N, Srikanth S, Sameer Kumar CV, Srivastava R, Saxena RK, Varshney RK (2014) In: M. Singh et al. (Eds) Broadening the genetic base of Grain legumes, Springer, India Edition 1, pp 149–159. doi: 10.1007/978-81-322-2023-7_7

Mir RR, Kudapa H, Srikanth S, Saxena RK, Sharma A, Azam S, Saxena K, Penmetsa RV, Varshney RK (2014) Candidate gene analysis for determinacy in pigeonpea (Cajanus spp.). Theor Appl Genet 127 (12):2663–2678

Moraes AP, Zanettini MHB, Jacques SMC, Santos EK (2004) Effect of temperature shock on soybean microspore embryogenesis. Braz Arch Biol Technol 47(4):537–544

Murthy KSR, Emmannuel S (2011) Nutritional and anti-nutritional properties of the unexploited wild legume Rhynchosia bracteata benth. Bangadesh J Sci Ind Res 46:141–146

Nair S, Prasada Rao U, Bennett J, Mohan M (1995) Detection of a highly heterozygous locus in recombinant inbred lines of rice and its possible involvement in heterosis. Theor Appl Genet 91(6):978–986

Ochatt S, Pech C, Grewal R, Conreux C, Lulsdorf M, Jacas L (2009) Abiotic stress enhances androgenesis from isolated microspores of some legume species (Fabaceae). J Plant Physiol 166:1314–1328

Oke DB, Tewe OO, Fetuga BL (1995) The nutrient composition of some cowpea varieties. Nigerian J Anim Prod 22:32–36

Peng J, Korol AB, Fahima T, Rodert MS, Ronin YI, Li YC, Nevo E (2000) Molecular genetic maps in wild emmer wheat, Triticum dicoccoides: genome-wide coverage, massive negative interference and putative quasi-linkage. Genome Res 10:1509–1531

Pumphrey M, Bai J, Laudencia-Chingcuanco D, Gill BS (2009) Non-additive expression of homeologous genes is established upon polyploidization in hexaploid wheat. Genetics 181:1147–1157

Pundir RPS, Singh RB (1985) Crossability relationships among Cajanus, Atylosia and Rhynchosia species and detection of crossing barriers. Euphytica 34:303–308

Punguluri SK, Janaiah K, Govil JN, Kumar PA, Sharma PC (2007) AFLP finger printing in pigeonpea (Cajanus cajan (L.) Mill sp.) and its wild relatives. Genet Resour Crop Evol 53:423–431

Rapp RA, Udall JA, Wendel JF (2009) genomic expression dominance in allopolyploids. BMC Biol 7:18. doi:10.1186/1741-7007-18

Remanandan P (1990) Pigeonpea: genetic resources. In The Pigeonpea (Nene YL, Hall SD, Sheila VK eds) Wallingford, 89–115, Oxon, UK: CAB International

Sateesh Kumar P (1985) Crossability, genome relationships and inheritance studies in intergeneric hybrids of pigeonpea. PhD Thesis, University of Hyderabad, Hyderabad, India

Saxena KB, Faris DG, Kumar RV (1987) Relationship between seed size and protein content in newly developed high protein lines of pigeonpea. Plant Food Hum Nutr 36:335–340

Saxena KB, Ariyanagayam RP, Reddy LJ (1992) Genetics of high selfing trait in pigeonpea. Euphytica 59:125–127

Saxena KB, Tikle AN, Kumar RV, Choudhary AK, Bahadur B (2016) Nectivore-aided hybridization and its exploitation for productivity enhancement in pigeonpea. Int J Sci Res Pub 6(8): 321–328

Singh NB, Singh IP, Singh BB (2005) Pigeonpea breeding. Advances in pigeonpea research. Indian Institute of Pulses Research, Kanpur, pp 67–95

Srikanth S, Rao MV, Mallikarjuna N (2013) Interspecific hybridization between Cajanus cajan (L.) Millsp. and C. lanceolatus (WV Fitgz) van der Maesen. Plant Genet Resour 12(2):255–558

Srikanth S, Marri S, Kollipara P, Rao MV, Mallikarjuna N (2017) Protease inhibitors of Cajanus conferring resistance to pod borer of pigeonpea (Cajanus cajan L. Millsp). Electronic J Plant Breed 8(1): 29–37

Srikanth S, Saxena RK, Rao MV, Varshney RK, Mallikarjuna N (2015) Development of a new CMS system in pigeonpea utilizing crosses with Cajanus lanceolatus (WV Fitgz) van der Maesen. Euphytica 204(2):289–302

Upadhyaya HD, Reddy LJ, Gowda CLL, Reddy KN, Singh S (2006) Development of a mini core subset for enhanced and diversifi ed utilization of pigeonpea germplasm resources. Crop Sci 46:2127–2132

van der Maesen LJG (1980) India is the native home of pigeonpea. In: Arends JC, Boelema G, de Groot CT, Leeuwenberg AJM (Eds) Ibergratulatorius in honorem H.C.D. de Wit landbouwhoge school miscellaneous paper no 19. Veenman H, Zonen BV, Wageningen, pp 257–262

van der Maesen LJG (1986) Cajanus DC and Alylosia W. & A (Leguminosae). Agricultural University Wageningen papers 85-4 (1985). Agricultural University Wageningen, the Netherlands, p 222

Wang J, Tian L, Lee HS, Chen ZJ (2006a) Non-additive regulation of FRI and FLC loci mediates flowering time variation in Arabidiopsis allopolyploids. Genetics 173:965–974

Wang J, Tian L, Lee HS et al (2006b) genome wide non-additive gene regulation in Arabidiopsis allotetraploids. Genetics 172:507–517

Wright B (1997) Crop genetic resource policy: the role of ex situ gene banks. Aust J Agric Resour Econ 41: 87–115

Modern Genomic Tools for Pigeonpea Improvement: Status and Prospects

5

Abhishek Bohra, Shalini Pareek, Rintu Jha, Rachit K. Saxena, Indra P. Singh, Gaurav Pandey, Raj K. Mishra, Farindra Singh, Mayank Kaashyap, Rohit Joshi and Rajeev K. Varshney

Abstract

Pigeonpea owing to its ability to sustain harsh environment and limited input/water requirement remains an excellent remunerative crop in the face of increasing climatic adversities. With nearly 70% share in global pigeonpea production, India is the leading pigeonpea producing country. Since the mid-1900s, constant research efforts directed to improve yield and resistance levels of pigeonpea have resulted in the development and deployment of several commercially accepted cultivars in India, accommodating into diverse agro-climatic zones. However, the crop productivity needs incremental improvements in order to meet the growing nutritional demands, especially in developing countries like India where pigeonpea forms a dominant part of vegetarian diet. Empowering crop improvement strategies with genomic tool kit is imperative to attain the project gains in crop yield. In the context, adoption of next-generation sequencing (NGS) technology has helped establish a wide range of genomic resources to support pigeonpea breeding, and the existing molecular tool kit includes genome-wide genetic markers, transcriptome/genome assemblies, and candidate genes/QTLs for target traits. Similarly, availability of whole mitochondrial genome sequence and derived DNA markers is immensely relevant in order to furthering the understanding of cytoplasmic male sterility (CMS) system and hybrid breeding. This chapter covers the progress of developing modern genomic resources in pigeonpea and highlights their vital role in designing future crop breeding schemes.

A. Bohra (✉) · S. Pareek · R. Jha · I.P. Singh · G. Pandey · R.K. Mishra · F. Singh
Crop Improvement Division, Indian Institute of Pulses Research (IIPR), Kanpur 208024, India
e-mail: abhi.omics@gmail.com

R.K. Saxena · R.K. Varshney
International Crops Research Institute for the Semi-Arid Tropics (ICRISAT), Patancheru 502324, India

M. Kaashyap
School of Applied Sciences, Health Innovations Research Institute, RMIT University, Melbourne, VIC 3000, Australia

R. Joshi
Plant Stress Biology, International Centre for Genetic Engineering and Biotechnology, New Delhi 110067, India

R.K. Varshney et al. (eds.), *The Pigeonpea Genome*,
Compendium of Plant Genomes, DOI 10.1007/978-3-319-63797-6_5

5.1 Introduction

Pigeonpea [*Cajanus cajan* (L.) Millspaugh] is an important food legume crop of the semi-arid tropics (SATs). Its ability to withstand risk-prone environments and low-input conditions makes it a preferred crop for the farming community relying on subsistence agriculture (Varshney et al. 2012). Breeding through conventional means (selection and hybridization) has been fruitful, and more than 100 varieties belonging to different maturity groups have been released for commercial cultivation in India during the last 50 years (Singh et al. 2016a). These varieties adequately suit to the specific needs of the diverse agro-climatic zones. However, pigeonpea productivity that has stayed consistently low (around 700–800 kg/ha) over past several decades remains far below from the demand in global food production projected by 2050. A 70% increase in food production worldwide necessitates transformative changes in methods used for crop breeding and management (Tester and Langridge 2010). Implementation of modern genomic resources holds great promise to attend the challenge mentioned above. Collaborative research efforts have facilitated development of much needed genomic tools for pigeonpea improvement during the last ten years (Varshney et al. 2013; Bohra et al. 2014; Pazhamala et al. 2015). Like other crops, dramatic impact of evolution in sequencing chemistry was evident in pigeonpea. The modern tools established recently in pigeonpea include sequence-based molecular markers and high-density genotyping/sequencing assays, saturated genome maps and comprehensive transcript assemblies, and most importantly, candidate gene(s)/QTLs for important traits. *De novo* sequencing attempts rendered entire genome sequence of a leading variety 'Asha' available for future research and breeding.

5.2 Economic Significance and Production Constraints

Pigeonpea is largely cultivated under rainfed conditions predominantly as an intercrop with other crops like sorghum, maize, cotton, soybean, and sunflower (Sameer Kumar et al. 2016). India is the largest producer of pigeonpea contributing 3.29 Mt of the total 4.85 Mt harvested worldwide followed by Myanmar (0.57 Mt), Malawi (0.30 Mt), and Kenya (0.27 Mt) (FAOSTAT 2014). Most of the Indian population depends on pigeonpea for protein source either as split dal or as vegetable along with cereals for balanced diet (Sharma et al. 2011; Kabuo et al. 2015). According to the UN report, Indian population is expected to reach 1.69 billion in the year of 2050. Therefore, pigeonpea remains an important crop with regard to providing food and nutritional security to a large segment of India population (Abraham et al. 2014; Saxena et al. 2015).

The productivity of this crop is severely hampered by a variety of diseases (Reddy et al. 1998) and insect pests (Sharma et al. 2010). Sterility mosaic disease (SMD) presents serious threat to pigeonpea production, and up to 95% yield loss has been registered in SMD-affected pigeonpea (Reddy and Nene 1981). Kannaiyan and Nene (1981) reported that *Fusarium* wilt (FW) caused by *Fusarium udum* in pigeonpea influences different stages (from pre-pod to pre-harvest) of the crop, causing 30 to 99% yield loss. Similarly, *Phytophthora* blight of pigeonpea caused by *Phytophthora drechsleri* f. sp. *cajani* leads to potential economic loss in pigeonpea (Pande et al. 2011). Among insect pests, pod borers are reported to have devastating impact on pigeonpea production, and pod fly (*Melanagromyza obtusa*) is another important biotic factor that challenges pigeonpea cultivation (Sharma et al. 2011).

Weak drainage system and water stagnation exert pronounced impact on pigeonpea yield (Reddy 2009). Drought and waterlogged conditions remain crucial abiotic factors that constrain pigeonpea production (Chauhan 1990). As reported by Kumutha et al. (2008), pigeonpea plants exhibit severe loss in relative water content (RWC) and chlorophyll of leaves and membrane stability index (MSI) in both roots and leaves under water-logged conditions. Based on the survey of pigeonpea grown across several locations in Kenya, substantial decrease in pigeonpea production might be credited to a variety of abiotic stresses, losses even extending up to 100% (Mergeai et al. 2001). Similarly, Mehta and Srivastava (2000) reported considerably higher reduction in pigeonpea production in India during drought years.

5.3 Taxonomy and Cytogenetics

Pigeonpea belongs to the family Fabaceae (subfamily: papilionaceae) under the tribe Phaseoleae (see Bohra et al. 2010). Divergent views were offered regarding the origin of pigeonpea with some authors favoring India (Vavilov 1951) while others advocating for Africa (Zeven and Zhukovsky 1975). On the basis of crop diversity, van der Maesen (1980) definitively suggested that India should be primary center of origin and Africa may be secondary center of origin. The close proximity of *C. cajan* and *C. cajanifolius* has been established through a wide range of diversity studies involving various wild species, which implies toward latter being the most probable progenitor of cultivated pigeonpea (van der Maesen 1980, 1990; Kassa et al. 2012). Similarly, karyotypic features including the morphology of chromosomes and the number of satellite chromosomes were reported to be strikingly similar between *C. cajan* and *C. cajanifolius* (Ohri and Singh 2002). Similar to the observation made by Roy (1933) and Naithani (1941) regarding chromosome count in pigeonpea, analysis of somatic chromosomes of ten different species including *Cajanus*, *Atylosia,* and *Rhynchosia* led authors to report 11 pairs of chromosomes in pigeonpea genome with nearly symmetric karyotypes except of *Atylosia albicans* (Pundir and Singh 1986). No significant variation in genome sizes was reported within cultivated pigeonpea, i.e., *C cajan* on the basis of flow cytometry and Feulgen densitometry (Greilhuber and Obermayer 1998).

5.4 Genomic Tools in Pigeonpea

Advances in pigeonpea genomics led a dramatic expansion in the arsenal of genomic resources that are greatly relevant to breeding (Table 5.1). In this section, we offer a brief account on these modern genomic tools and technologies.

5.5 Next-Generation Mapping Resources

A genetic population of moderate size segregating for the desired trait(s) is essentially needed to find significant associations between the DNA markers and trait(s) under consideration. Experimental populations stemming from a cross between two contrasting genotypes have been developed in pigeonpea targeting several traits such as resistance to important biotic/abiotic factors, fertility restoration, and growth habit/flowering patterns (Varshney et al. 2010; Khalekar et al. 2014; Pazhamala et al. 2015; Daspute and Fakrudin 2015). In parallel, reverse genetic tools like targeted induced local lesions in genomes (TILLING) population derived from EMS-treated 'Asha' were also reported in pigeonpea (Varshney et al. 2010). The reference mapping population in pigeonpea was generated from an interspecific cross [ICP 28 (*C. cajan*) ICPW 94 (*C. scarabaeoides*)], which eventually served for the development of reference linkage maps of moderate (SSR based) to high density (SNP based). Concomitant with the availability of high-density marker assays, a conceptual shift has been witnessed in designing the mating schemes that has paved way for the establishment of modern mapping resources involving a

Table 5.1 Genomic resources in pigeonpea

Resource		Reference
High-throughput genotyping assays	GoldenGate	Kumawat et al. (2012)
	KASP	Saxena et al. (2012, 2014)
	VeraCode	Roorkiwal et al. (2013)
Modern genetic populations	MAGIC	see Pazhamala et al. (2015)
	NAM	see Pazhamala et al. (2015)
High-density genome map	910 loci (interspecific F2)	Saxena et al. (2012)
Large-scale genetic variants	SNP	Dubey et al. (2011), Singh et al. (2011), Varshney et al. (2012), Saxena et al. (2012)
	SSR	Bohra et al. (2011), Singh et al. (2011), Varshney et al. (2012)
Marker-trait associations (MTAs)	Fertility restoration (SSR)	Bohra et al. (2012)
	Fusarium wilt (SSR/SNP)	Khalekar et al. (2014), Singh et al. (2015)
	SMD (SSR/SNP)	Gnanesh et al. (2011a, 2011b), Singh et al. (2015)
	Plant type and earliness (SSR/SNP)	Kumawat et al. (2012), Geddam et al. (2014)
	Flowering pattern/determinacy (SNP)	Mir et al. (2014)
Transcriptome assemblies	4557 TACs	Raju et al. (2010)
	43324 TACs	Dutta et al. (2011)
	48726 TACs	Dubey et al. (2011)
	21434 TACs	Kudapa et al. (2012)
Mitochondrial genome sequence	545.7 Kb	Tuteja et al. (2013)
Whole genome sequences	510.8 Mb	Singh et al. (2011)
	605.7 Mb	Varshney et al. (2012)
Molecular assays to assist CMS breeding	21 SSRs	Saxena et al. (2010a, 2010b)
	42 SSRs	Bohra et al. (2011)

set of diverse founders (Bohra 2013). The two widely employed mating designs incorporating multiple parents are multi-parent advanced-generation intercross (MAGIC) and nested association mapping (NAM). These multi-parental populations are being increasingly reported across several crops like maize (McMullen et al. 2009), wheat (Huang et al. 2012; Delhaize et al. 2015), rice (Bandillo et al. 2013), pea (Tayeh et al. 2015), sorghum (Ongom et al. 2016). Availability of a reference genome sequence for 'Asha' genotype encouraged its use as a common parent in NAM scheme. Such new-generation mapping populations not only ensure the best utilization of high-throughput genotyping/sequencing platforms but also offer several advantages over conventional (biparental) mapping populations like greater resolution, allelic richness. Also, these adequately address the caveats associated with the association analysis (AA) such as the inflated rate of false positives. These advantages render these mapping populations suitable for both family-based QTL study and AA or more appropriately, the joint linkage—linkage disequilibrium (LD) analysis.

5.6 Genome-Scale DNA Markers

Diverse marker assays were employed in pigeonpea for a variety of purposes including assessment of genetic diversity, linkage mapping, and QTL analyses. The first generation of markers included restriction fragment length polymorphisms (RFLPs: Nadimpalli et al. 1992), random amplified polymorphic DNA (RAPDs: Ratnaparkhe et al. 1995) followed by the DNA markers of second generation such as simple sequence repeats (SSRs). Initially, SSR markers were developed from genomic libraries using conventional protocols (Odeny et al. 2007; Saxena et al. 2010a). A limited throughput coupled with the higher cost of the marker development and subsequent genotyping assays urgently called for DNA marker systems that are amenable in terms of throughput, cost, and accuracy. The first set of massive DNA markers in pigeonpea was reported by Bohra et al. (2011). The authors developed more than 3000 SSRs from BAC-end sequences (BESs) and successfully applied these markers in linkage analysis, hybridity testing, and other genetic analyses. The new generation of markers including diversity arrays technology (DArT) and single nucleotide polymorphism (SNP) markers extended marker coverage to genome level. The DArT assays in pigeonpea enabled assessment of the genetic variation and linkage mapping. Among the several marker systems advancing contemporarily, SNP is increasingly replacing SSRs as the DNA marker of choice. A set of 1616 SNPs designated as pigeonpea KASP assay markers (PKAMs) was subsequently used to analyze a panel of 24 genotypes and to construct a high-density linkage map (Saxena et al. 2012). Further, a subset of these PKAMs was selected on the basis of polymorphism among cultivated types, polymorphism information content (PIC) values, and assay design tool (ADT) scores, and 256 genotypes of pigeonpea reference set were analyzed using 48-plex VeraCode Assay technology on the BeadXpress platform (Roorkiwal et al. 2013). This represented an important study concerning the assessment of genetic diversity that holds greater relevance to breeder community. The 1,616 SNPs were also used to screen 184 *Cajanus* accessions (77 cultivated and 107 wild relatives from secondary and tertiary gene pool), which led to the identification of a greater number of polymorphic DNA markers (1226). Importantly, greater insights into domestication of pigeonpea were gained supporting the long-established view that *C. cajanifolius* is the closest progenitor and Madhya Pradesh is the center of origin (Saxena et al. 2014). In parallel, whole transcriptome and genome assemblies also served for the identification of large-scale DNA markers. Genetic variations were reported in the form of expressed sequenced tag (EST)—SSRs, intron spanning region (ISR) markers, and SNPs via excavating transcriptome assemblies (Raju et al. 2010; Dutta et al. 2011; Dubey et al. 2011; Kudapa et al. 2012). Likewise, genome-wide SSRs and SNPs were also recovered from whole genome sequence of pigeonpea. Increasing access to such high-throughput and cost-effective marker systems will certainly help improving the efficiency of traditional breeding methods.

5.7 Molecular Linkage Maps

No genetic linkage map was reported in pigeonpea till 2011, and this lack of linkage information might be credited largely to the inadequacy of polymorphic DNA markers and the lack of mapping populations. Access to the modern marker technology like DArT enabled development of first genetic map in pigeonpea for an interspecific cross (*C. cajan* × *C. scarabaeoides*). However, the paternal and maternal specific maps could not be integrated into a single genetic map, thereby restricting its widespread use in future research work. The first SSR-based genetic map comprising 239 loci was reported for the same interspecific cross that spanned 930 cM of the pigeonpea genome (Bohra et al. 2011). In a similar fashion, genetic linkage maps were developed for cultivated crosses as well, in which the number of mapped SSR loci ranged from 59 to 140 (Gnanesh et al. 2011a, 2011b; Bohra et al. 2012). A 296-loci (genic SNP and SSR) genetic map for cultivated pigeonpea was constructed by Kumawat et al. (2012) covering 1520 cM of the genome. Apart from these population-specific maps, the first consensus genetic map with 339 loci was synthesized by integrating marker information from six different F_2 populations (Bohra et al. 2012). Extremely low DNA polymorphism as revealed by SSRs or other earlier prevailing DNA marker systems demanded a shift toward adoption of high-throughput marker technologies such as genome-wide SNPs, and as a result of SNP markers assayed through KASP platform, a saturated genetic map was obtained for an interspecific F_2 population (*C. cajan* × *C. scarabaeoides*). The map covered a map distance of 996 cM with 910 (SNPs and SSRs) spaced at an average marker distance of 1.09 cM (Saxena et al. 2012).

5.8 Comprehensive EST Resources and Transcriptome Assemblies

Prior to the introduction of NGS techniques in pigeonpea, different research groups (Raju et al. 2010, Priyanka et al. 2010; Kumar et al. 2014) used Sanger-derived EST resources to access the transcribed regions in the pigeonpea genome. With the aim to develop and analyze ESTs responsive to FW and SMD, Raju et al. (2010) generated the first set of ESTs for maker development in pigeonpea. A total of 9,468 high-quality ESTs were obtained through sequencing 16 cDNA libraries of four pigeonpea genotypes that respond to FW (ICPL 20102 and ICP 2376) and SMD (ICP 7035 and TTB 7). The authors also found 19 and 20 genes to be differentially expressed, respectively, in FW- and SMD-responsive genotypes. Similarly, pigeonpea ESTs were characterized and the genes responsive to abiotic stress were functionally validated in *Arabidopsis thaliana* (Priyanka et al. 2010). From the cDNA libraries of drought stressed pigeonpea, 75 high-quality ESTs were obtained, of which 20 were specific to pigeonpea. Overexpression of three selected pigeonpea genes, viz. *CcHyPRP* (*Cajanus cajan* hybrid-proline-rich protein), *CcCYP* (*C. cajan* cyclophilin), and *CcCDR* (*C. cajan* cold and drought regulatory) genes, in *Arabidopsis* confirmed the plant's tolerance under abiotic stress. Kumar et al. (2014) generated 105 high-quality ESTs from the root tissues of pigeonpea genotype GRG295. Further, the expression of four transcripts, namely *S-adenosylmethionine synthetase*, *phosphoglycerate kinase*, *serine carboxypeptidase*, and *methionine aminopeptidase,* was validated through reverse transcriptase PCR.

Dutta et al. (2011) sequenced transcriptomes of two pigeonpea genotypes 'Asha' and 'UPAS 120' using 454 GS-FLX pyrosequencing. The total number of transcript assembly contigs (TACs) was 43,324. Further analysis of this assembly captured more than 3,000 genic SSR markers. Moreover, primer pairs could be designed for 2,877 SSRs, and 550 (designated as ASSRs) of these SSRs provided the amplicons of expected size. Another assembly '*Cajanus cajan* transcriptome assembly' (CcTA v1) was developed by combining 454-derived 494,353 short transcript reads (STRs) for Pusa Ageti (ICP 28) with 10,817 Sanger ESTs. The assembly comprised of 48,726 (38.1%) contigs and 79,028 singletons, and N50 of this assembly was 287 bp. A search for differentially expressed TUSs resulted in the identification of 99 and 13 TUSs common to FW- and SMD- responsive genotypes, respectively. Moreover, a set of 8,137 SSRs was extracted from this assembly and a total of 12,141 SNPs were detected across different parental combinations. The most comprehensive assembly (CcTA v2) was reported recently by Kudapa et al. (2012) by combining 18,353 Sanger ESTs with reads generated from different sequencing platforms such as Illumina (128.9 million reads) and FLX/454 (2.19 million reads). The assembly composed of total of 21,434 TACs with N50 of 1,510 bp. The transcript reads assembled in CcTA v2 were generated from 16 different pigeonpea genotypes. A comparison of this assembly with soybean genome sequence permitted discovery of 10,009 ISR markers, and a subset of 116 yielding scorable amplicons was used to screen eight pigeonpea genotypes. Together with enabling access to the functionally important segments of the pigeonpea genome, these transcriptomic tools represent an important community resource to facilitate comparative genomics and offer transferable DNA markers for cross-genera studies.

5.9 Reference Genome Sequence

Two whole genome assemblies have been reported in pigeonpea for the genotype 'Asha' by two different research groups. Varshney et al. (2012) assembled more than 70% of the entire 833 Mb genome using Illumina sequencing platform. This assembly revealed the presence of a total of 48,680 genes with a mean transcript length of 2,348.70 bp. Like other sequenced legume crops, nearly half of the pigeonpea genome was reported to contain repetitive elements (REs). New light was shed on the genetic landscape of drought tolerance in pigeonpea with detection of 111 candidate genes. Serving as a massive reservoir of the genetic markers, this assembly delivered large sets of SSRs (23,410 primer pairs synthesized of total 309,052) and SNPs (28,104). In a similar manner, Singh et al. (2011) assembled 510 Mb (nearly 60%) of the pigeonpea genome with the help of 454 sequencing system. The number of protein-coding genes in this assembly was similar to what was reported by Varshney et al. (2012); however, the average gene size was reported to be 1,170 bp. Among the total 47,004 genes contained in the genome, 1,213 were noted to be disease/defense responsive, whereas 152 genes were predicted to regulate plant's response to abiotic stress. Establishment of a reference genome sequence improves scope for future resequencing attempts and other genome-wide mapping methods including next-generation mapping (NGM). Further, decoding of the entire genome sequence of a leading pigeonpea variety will greatly assist breeders and geneticists to develop improved cultivars or hybrids, particularly to accommodate in a climate increasingly challenged by biotic and abiotic constraints. Further, coupling traditional breeding techniques to modern omics technologies will help ensure a promising future to the pigeonpea farmers and economy.

5.10 Progress Toward Finding Candidate Genes/QTL(S) Related to Target Traits

Determination of the causative gene(s)/genomic segments that explain considerable proportion of the phenotypic variation (PV) for any given trait forms a crucial step in genomics-assisted crop improvement. Examination of marker-trait associations (MTAs) entails the generation of experimental populations or an existing panel of diversified genotypes. The former is termed as family-based linkage (FBL) mapping, while the latter is known as Association Analysis (AA) (Mackay and Powell 2007).

5.11 Genetic Inheritance and Gene/QTL Discovery

As described in earlier sections, FW and SMD are the major diseases that raise tremendous concerns among the pigeonpea growers. Concerning the genetic inheritance, resistance to these two important diseases was reported to be governed by recessive gene(s) with varying numbers (Odeny et al. 2009; Jain and Reddy 1995; Gnanesh et al. 2011a). However, contradictory reports showing resistance conferred by one/two dominant genes are also available (Murugesan et al. 1997, Singh et al. 2016b). In case of FW, complementary and inhibitory gene actions were noted (Ajay et al. 2013), while duplicate dominant epistasis and inhibitory epistasis (Daspute et al. 2014) are reported to play important roles in case of SMD resistance.

Several biparental mapping populations have been reported in pigeonpea that segregated for different traits. Two methods, i.e., bulked segregant analysis (BSA) and QTL mapping, were used for mapping these traits in pigeonpea. Examples of F_2-based BSA include RAPD markers for FW (Kotresh et al. 2006) and plant type (Dhanasekar et al. 2010), and AFLP marker for SMD

(Ganapathy et al. 2009) (see Bohra et al. 2014). More recent application of BSA in F_2 population (Gullyal white × BSMR 736) facilitated detection of a repulsive-phase RAPD fragment co-segregating with SMD resistance (Daspute and Fakrudin 2015). Similarly, Khalekar et al. (2014) found five SSR markers to be associated with FW resistance by analyzing resistant and susceptible bulks of an F_2 population (ICPL 87119 × T. Vishakha-1).

In addition, QTL analysis in F_2 populations unearthed putative genomic segments controlling a considerable proportion of the PV. For instance, Gnanesh et al. (2011b) analyzed F_2 population and $F_{2:3}$ families (ICP 8863 × ICPL 20097 and TTB 7 × ICP 7035) for SMD isolates (Patancheru and Bengaluru) and detected QTLs having minor as well as major effects on the phenotype. The PV ranged between 8.3 and 24.7%. Similarly, QTL analysis performed on three F_2 populations segregating for male sterility and fertility enabled discovery of four major effect QTLs explaining PV in the range of 14.85–24.17%. Likewise, Kumawat et al. (2012) discovered QTLs for plant type and earliness from the $F_{2:3}$ population (Pusa Dwarf × HDM04-1). Of these QTLs, one could explain PV up to 51.4%. Interestingly, these QTLs were later validated in the recombinant inbred background of the population 'Pusa Dwarf × H2001-4' (Geddam et al. 2014). For instance, qSB 5.1 was found explaining 15.1% of the PV in this study (Geddam et al. 2014), while the same QTL accounted for 10.4% PV in the previous report (Kumawat et al. 2012). Once validated in different genetic backgrounds, these putative genomic regions/associated DNA markers can be immediately deployed in crop improvement schemes.

5.12 Association Analysis (AA) or Linkage Disequilibrium (LD) Mapping

Alternatively, significant MTAs are discovered using association panel, which includes genetically diverse genotypes such as elite cultivars, landraces, or wild relatives. Compared to family-based (in particular biparental) analysis, the ability of AA or LD mapping to harness historical recombination makes it more efficient strategy in terms of allelic richness and mapping resolution. Equally important, no additional efforts are needed for generating experimental material and favorably, phenotypic data are often available on the diversified panel. LD analysis can be performed at pathway level using candidate genes or at whole genome scale (genome-wide association study: GWAS). In pigeonpea, a diversity panel comprising 94 lines representing mapping parents and germplasm lines could constitute a representative panel for determinacy trait as 11 of these exhibit a determinate (DT) growth habit while remaining 83 had indeterminate (IDT) habit. A genome-scale search for MTAs was made on these 94 lines using 6,144 DArTs and 768 GoldenGate SNPs. A total of 25 markers including 6 DArTs and 19 SNPs were detected as significantly associated with the trait and the phenotypic variance (R^2) explained by these markers reached up to 14.5% (Mir et al. 2012). Later, a candidate gene-based approach was adopted to decipher the genetic mechanism that controls determinacy trait in pigeonpea. A total of 142 genotypes were investigated using seven genes specific to determinacy/flowering pattern. The study resulted in the establishment of (*Cajanus cajan* terminal flower 1) *CcTFL1* as the likely candidate gene underlying determinacy/flowering pattern. Notably, the same candidate gene was identified through single marker analysis (SMA) and composite interval mapping (CIM) analyses in an F_2 mapping population (ICPA 2039 × ICPR 2447). SIM analysis revealed that *CcTFL1* accounted for 75% PV whereas as revealed by CIM the marker interval *CcTFL1*-CcM0126 was found to control up to 90% PV for determinacy trait. A qRT–PCR experiment performed on the two contrasting genotypes, i.e., ICPA 2039 (DT) and ICPL 87118 (IDT), further validated the experimental findings (Mir et al. 2014). To make the associated SNPs user-friendly, these markers were subsequently converted to PCR-amenable assays.

5.13 Reference Genes to Predict Response Under Abiotic Stresses

More recently, attempts were made to find a set of 'reference' genes to support analysis of gene expression in pigeonpea under the conditions challenged by abiotic stresses, in particular drought, heat, and salt stress (Sinha et al. 2015a, 2015b). Expression variation was measured for ten pigeonpea-specific genes orthologous to commonly used housekeeping genes (*EF1α, UBQ10, GAPDH, 18SrRNA, 25Sr RNA, TUB6, ACT1, IF4α, UBC,* and *HSP90*). The root, stem, and leaf tissues of the popular genotype 'ICPL 87119' were used for the analysis. As a result of quantitative assessment of gene expression, stable genes were obtained with regard to drought (IF4α and TUB6: Sinha et al. 2015a), heat (*UBC, HSP90, GAPDH*: Sinha et al. 2015b), and salt (*GAPDH, UBC, HSP90*: Sinha et al. 2015b) stress, which could act as internal control for expression studies in pigeonpea, as specific reference genes are required for specific species for given stress conditions.

5.14 Next-Generation Trait Mapping

Unlike the conventional mapping approaches involving multiple steps, analysis of mapping population with NGS facilitates discovery and mapping of genetic variants in an instantaneous fashion, and importantly, nucleotide level resolution can be achieved (Schneeberger 2014). In the context, various strategies have been proposed in recent years that effectively integrate high-throughput genotyping/sequencing protocols into mapping schemes (Varshney et al. 2013, Bohra and Singh 2015). In pigeonpea, NGS-based QTL Seq of pooled samples (susceptible and resistant extremes) of a recombinant inbred population (ICPL 20096 × ICPL 332) coupled with the resequencing data generated for four genotypes ICPL 20097, ICP 8863, ICPB 2049, and ICPL 99050 enabled discovery of four non synonymous (ns) SNPs each for FW (four candidate genes) and SMD (three candidate genes). Furthermore, expression study using qRT-PCR helped substantiate the robustness of the candidate genes that condition resistance against these two important diseases, i.e., '*C.cajan*_03203' for FW and '*C.cajan*_01839' for SMD (Singh et al. 2015).

5.15 Genomics to Underpin Cytoplasmic Genetic Male Sterility (CMS)-based Hybrid Breeding

Heterosis breeding constitutes an important approach for pigeonpea genetic improvement. Discovery of a stable CMS system in pigeonpea has offered time-saving and cost-effective ways to harness hybrid vigor. As reviewed recently by different researchers (Saxena et al. 2015; Bohra et al. 2016), deployment of genomic tools can help improving efficiency of CMS-based crop breeding programmes.

5.16 Mitochondrial Genome Sequence and Derived Molecular Tools

CMS offers a unique opportunity not only to understand the cytoplasm and nucleus interaction but also to breed hybrids with improved performance. CGMS is a maternally inherited trait, and the factors conditioning male sterility are known to reside within the mitochondrion. These male sterility-inducing causative genes or open reading frames (*orfs*) have been identified in various crops in which CMS system exits (Touzet and Meyer 2014; Horn et al. 2014; Hu et al. 2014). In order to identify these causative *orfs*, sequencing the whole genome of the mitochondrion remains an attractive strategy. In pigeonpea, mitochondrial genome sequence was established with master circle covering length of 545 Kb (Tuteja et al. 2013). The mitochondrial genome harbored a total of 51 genes, of which 34 were found to be protein coding. Sequence comparison of mitochondrial genomes between A and B lines facilitated detection of 13 chimeric *orfs*, which could be considered as putative genomic candidates inducing CMS in pigeonpea. These chimeric *orfs* are outcome of extensive genomic rearrangement events that occur in mitochondria. However, the exact genomic segment causing CMS remains to be pinpointed. An elaborated study involving interaction of both *Rf* and *orf* is warranted in order to generate greater insights into CGMS phenomenon. More recently, examination of expression profiling of 34 protein-coding genes from mitochondria led to the identification of nine differentially expressed genes, of which *nad4L* and *nad7a* also showed SNP and InDel, respectively. A PCR-based marker developed by targeting *nad7a* would prove helpful in identification between A and B lines (Sinha et al. 2015c). Survey of this mitochondrial genome further provided a set of 24 SSR markers (Khera et al. 2015).

5.17 Molecular Assays for Genetic Purity Assessment

In addition to mitochondrial-specific DNA markers, SSR marker from genomic libraries and BESs were used in CMS breeding schemes. Through enabling the assessment of genetic purity of hybrids and their parents, these offer valuable tools to support procedures to perform grow out test (GoT). Based on the SSR analysis of hybrid (ICPH 2438) carrying A_4 cytoplasm and its parents (ICPA 2039 and ICPR 2438), a set of 21 SSR markers was found to be informative because these SSRs generated monomorphic fragments between ICPA 2039 and ICPB 2039, and at the same time, these markers successfully discriminated ICPA 2039 and ICPR 2438. Further, out of these 21 SSR markers, two SSRs (CCB4 and CCttc006) were found to be diagnostic while testing the genetic purity of CMS-based hybrid ICPH 2438. Similarly, a total of 42 SSR markers were identified for facilitating purity testing of two hybrids ICPH 2671 and ICPH 2438. Importantly, SSR protocols were optimized to accommodate up to eight SSR markers in a multiplex (Bohra et al. 2011). More recently, seven SSR markers (CCB9, HASSR3, HASSR9, HASSR23, HASSR35, HASSR37, and HASSR43) were obtained by screening an A_2-cytoplasm-derived hybrid IPH 09-5 and its parents (PA 163 A and AK 261322) with 66 hyper-variable and informative SSRs (Bohra et al. 2015). Likewise, assaying CMS line (GT 288 A), restorer (GTR 11), and derived hybrid (GTH 1) with 40 SSRs facilitated identification of one DNA marker (CcM0030), which could be informative while assessing the heterozygous nature of hybrid GTH 1 (Patel et al. 2012).

5.18 Prospects for Fast-Track Trait Introgression and Molecular Breeding

The genomics-enabled crop improvement is at its initial stage in pigeonpea. The progress in last 10 years has been satisfactory in terms of generation of valuable genomic tools. This period represents ‘development’ or ‘training’ phase of the molecular breeding in which important marker-trait associations (MTAs) are established for downstream selection procedures or prediction models are trained for genomic selection (GS) (Nakaya and Isobe 2012; Bohra 2013). The real potential of genomics-assisted breeding will start unleashing once we enter into the ‘breeding phase.’ For traits that lie under the control of major effect QTL/gene, marker-assisted backcrossing (MABC) will be the most appropriate strategy for defect elimination, thus precisely improving an otherwise elite cultivar for the trait under consideration. At the same time, advanced backcross (AB)-QTL offers exciting avenues for detection as well as transfer of the traits. Example includes advanced segregating generations derived from wide crosses involving *C. scarabaeoides* as the wild donor (Varshney et al. 2013). By its virtue, AB-QTL seeks unexploited wild gene(s)/allele(s) that are usually absent in cultivated gene pool. Further, implementation of genome-wide approaches like GWAS and MAGIC/NAM in light of the NGS advances is likely to expand the array of robust genomic segments associated with the trait along with guiding the community for prioritizing the candidate genes. Growing adoption of schemes like GS will help to curtail the cost and time invested in repeated phenotypic screening.

5.19 Conclusion

Nutrient-dense crops like pigeonpea remain important when viewed from the point of food security and subsistence farming. New scientific interventions are needed to be in place in order to meet the challenges that the current agriculture faces worldwide. In recent years, genomics-assisted breeding emerges as a promising approach for accelerating crop production per unit area. In the context, an attractive collection of genomic tools is now available to exercise genomics-assisted crop improvement. Importantly, CMS-based hybrid breeding technology will also be informed greatly by the current genomic developments.

Acknowledgements Authors acknowledge support from Indian Council of Agricultural Research (ICAR), New Delhi, India.

References

Abraham B, Araya H, Berhe T, Edwards S, Gujja B, Khadka RB (2014) The system of crop intensification: reports from the field on improving agricultural production, food security, and resilience to climate change for multiple crops. Agric Food Sec 3:4

Ajay BC, Prasad PS, Gowda MB et al (2013) Inheritance of resistance to Bangalore race of *Fusarium* wilt disease in pigeon pea (*Cajanus cajan* L.). Aust J Crop Sci 7:1520–1524

Bandillo N, Raghavan C, Muyco PA et al (2013) Multi-parent advanced generation inter-cross (MAGIC) populations in rice: progress and potential for genetics research and breeding. Rice 6:11

Bohra A (2013) Emerging paradigms in genomics-based crop improvement. Sci World J 2013:585467

Bohra A, Mallikarjuna N, Saxena KB, Upadhyaya H, Vales I, Varshney RK (2010) Harnessing the potential of crop wild relatives through genomics tools for pigeonpea improvement. J Plant Biol 37:83–98

Bohra A, Dubey A, Saxena RK, Penmetsa RV, Poornima KN, Kumar N et al (2011) Analysis of BAC-end sequences (BESs) and development of BES-SSR markers for genetic mapping and hybrid purity assessment in pigeonpea. BMC Plant Biol 11:56

Bohra A, Saxena RK, Gnanesh BN, Saxena KB, Byregowda M, Rathore A et al (2012) An intra-specific consensus genetic map of pigeonpea [Cajanus cajan (L.) Millspaugh] derived from six mapping populations. Theor Appl Genet 125:1325–1338

Bohra A, Saxena RK, Saxena KB, Sameerkumar CV, Varshney RK (2014). Advances in pigeonpea genomics. In: Gupta S, Nadarajan N, Sen Gupta D (Eds) Legumes in the Omic Era, New York, Heidelberg, London, Springer, pp 95–110

Bohra A, Singh NP (2015) Whole genome sequences in pulse crops: A global community resource to expedite translational genomics and knowledge-based crop improvement. Biotechnol Lett 37:1529–1539

Bohra A, Singh IP, Yadav AK, Pathak A, Soren KR, Chaturvedi SK et al (2015) Utility of informative SSR markers in the molecular characterization of cytoplasmic genetic male sterility-based hybrid and its parents in pigeonpea. Natl Acad Sci Lett 38:13–19

Bohra A, Jha UC, Adhimoolam P, Bisht D, Singh NP (2016) Cytoplasmic male sterility (CMS) in hybrid breeding in field crops. Plant Cell Rep 35:967–993

Chauhan YS (1990). Pigeonpea: optimum agronomic management. In: Nene YL, Hall S D, Sheil,VK (Eds) The pigeonpea, CAB International, Wallingford, U.K., pp 257–278

Daspute A, Fakrudin B (2015) Identification of coupling and repulsion phase DNA marker associated with an allele of a gene conferring host plant resistance to pigeonpea sterility mosaic virus (PPSMV) in pigeonpea (*Cajanus cajan* L. Millsp.). Plant Pathol J 31 (1):33–40

Daspute A, Fakrudin B, Bhairappanavar SB, Kavil SP, Narayana YD, Muniswamy A, Kaumar PU, Krishnaraj AY, Khadi BM (2014) Inheritance of pigeonpea sterility mosaic disease resistance in pigeonpea. Plant Pathol J 30:188–194

Delhaize E, Rathjen TM, Cavanagh CR (2015) The genetics of rhizosheath size in a multiparent mapping population of wheat. J Exp Bot 66:4527–4536

Dhanasekar P, Dhumal KH, Reddy KS (2010) Identification of RAPD marker linked to plant type gene in pigeonpea. Indian J Biotechnol 9:58–63

Dubey A, Farmer A, Schlueter J, Cannon SB, Abernathy B, Tuteja R et al (2011) Defining the transcriptome assembly and its use for genome dynamics and transcriptome profiling studies in pigeonpea (*Cajanus cajan* L.). DNA Res 18:153–164

Dutta S, Kumawat G, Singh BP, Gupta DK, Singh S, Dogra V et al (2011) Development of genic-SSR markers by deep transcriptome sequencing in pigeonpea [Cajanus cajan (L.) Millspaugh]. BMC Plant Biol 11:17

FAOSTAT (2014) FAOSTAT Database. Available online at: http://www.fao.org/faostat/en/#data/QC

Ganapathy KN, Byregowda M, Venkatesha SC, Rama Chandra R, Gnanesh BN, Girish G (2009)

Identification of AFLP markers linked to sterility mosaic disease in pigeonpea *Cajanus cajan* (L.) Millsp. Int J Integr Biol 7:145–149

Geddam SB, Raje RS, Prabhu KV, Singh NK, Chauhan DA, Jain P, et al (2014) Validation of QTLs for earliness and plant type traits in pigeonpea (Cajanus cajan (L.) Millsp.) Indian JGenet 74(4): 471–477

Gnanesh BN, Ganapathy KN, Ajay BC, Byre Gowda M (2011a) Inheritance of sterility mosaic disease resistance to Bangalore and Patancheru isolates in pigeonpea (*Cajanus cajan* (L.) Millsp.). Electron J Plant Breed 2:218–223

Gnanesh BN, Bohra A, Sharma M, Byregowda M, Pande S, Wesley V et al (2011b) Genetic mapping and quantitative trait locus analysis of resistance to sterility mosaic disease in pigeonpea [*Cajanus cajan* (L.) Millsp.]. Field Crops Res 123:53–61

Greilhuber J, Obermayer R (1998) Genome size variation in *Cajanus cajan* (Fabaceae): a reconsideration. Plant Syst Evol 212:135–141

Horn R, Gupta KJ, Colombo N (2014) Mitochondrion role in molecular basis of cytoplasmic male sterility. Mitochondrion 19:198–205

Hu J, Huang W, Huang Q, Qin X, Yu C, Wang L et al (2014) Mitochondria and cytoplasmic male sterility in plants. Mitochondrion 19:282–288

Huang BE, George AW, Forrest KL, Kilian A, Hayden MJ, Morell MK et al (2012) A multiparent advanced generation inter-cross population for genetic analysis in wheat. Plant Biotechnol J 10:826–839

Jain KC, Reddy MV (1995) Inheritance of resistance to *Fusarium* wilt in pigeonpea (*Cajanus cajan* (L) Millsp.). Ind J Genet Plant Breed 55:434–437

Kabuo NO, Dialoke SA, Onuegbu N, Nwosu JA, Peter-Ikechukwu AI, Ogbu IR (2015) Utilization of Tender Pigeonpea (*Cajanus cajan* (L.) Millsp.) in Nigeria. Food Sci Qual Manag 37:110–115

Kannaiyan J, Nene YL (1981) Influence of wilt at different growth stages on yield loss in pigeonpea. Trop Pest Manag 27:141

Kassa MT, Penmetsa RV, Carrasquilla-Garcia N, Sarma BK, Datta S, Upadhyaya HD et al (2012) Genetic patterns of domestication in pigeonpea (*Cajanus cajan* (L.) Millsp.) and wild *Cajanus* relatives. PLoS ONE 7(6): e39563

Khalekar GD, Akhare AA, Gahukar SJ, Singh NK, Kumar M (2014) Identification of simple sequence repeat markers associated with wilt resistance in pigeonpea. J Environ Biol 35(5):955–960

Khera P, Saxena R, Sameerkumar CV, Saxena K, Varshney RK (2015) SSRs and their utility in distinguishing wild species, CMS lines and maintainer lines in pigeonpea (*Cajanus cajan* L.). Euphytica. doi 10.1007/s10681-015-1504-2

Kotresh H, Fakrudin B, Punnuri S, Rajkumar B, Thudi M, Paramesh H et al (2006) Identifi cation of two RAPD markers genetically linked to a recessive allele of a *Fusarium* wilt resistance gene in pigeonpea (*Cajanus cajan* (L.) Millsp.). Euphytica 149:113–120

Kudapa H, Bharti AK, Cannon SB, Farmer AD, Mulaosmanovic B, Kramer R et al (2012) A comprehensive transcriptome assembly of pigeonpea (*Cajanaus cajan* L.) using Sanger and Second-generation sequencing platforms. Mol Plant 5:1020–1028

Kumar RR, Yadav S, Joshi S, Bhandare PP, Patil VK, Kulkarni PB et al (2014) Identification and validation of expressed sequence tags from pigeonpea (*Cajanus cajan* L.) root. Int J Plant Genom 2014:651912

Kumawat G, Raje RS, Bhutani S, Pal JK, Mithra SVCR, Kishor Gaikwad K et al (2012) Molecular mapping of QTLs for plant type and earliness traits in pigeonpea (*Cajanus cajan* L. Millsp.). BMC Genet 13:84

Kumutha D, Sairam RK, Ezhilmathi K, Chinnusamy V, Meena RC (2008) Effect of waterlogging on carbohydrate metabolism in pigeonpea (*Cajanus cajan* L.): Upregulation of sucrose synthase and alcohol dehydrogenase. Plant Sci 175:706–716

Mackay I, Powell W (2007) Methods for linkage disequilibrium mapping in crops. Trends Plant Sci 12:57–63

McMullen MM, Kresovich S, Villeda HS et al (2009) Genetic properties of the maize nested association mapping population. Science 325:737–740

Mehta R, Srivastava PK (2000) Analysis of seasonality in prices of agricultural commodities. Agric Situat India, September issue

Mergeai G, Kimani PM, Mwangombe A, Olubayo F, Smith C, Audi P et al (2001) Survey of pigeonpea production systems, utilisation and marketing in semi-arid lands of Kenya. Biotechnol Agron Soc Environ 5:145–153

Mir RR, Saxena RK, Saxena KB, Upadhyaya HD, Kilian A, Cook DR, Varshney RK (2012) Whole-genome scanning for mapping determinacy in pigeonpea (*Cajanus* spp.). Plant Breed 132:472–478

Mir RR, Kudapa H, Srikanth S, Saxena RK, Sharma A, Azam S et al (2014) Candidate gene analysis for determinacy in pigeonpea (*Cajanus* spp.). Theor Appl Genet 127:2663–2678

Murugesan S, Murugan E, Nadarajan N (1997) Inheritance of duration leaf colour sterility mosaic disease resistance and growth habit in pigeonpea. Madras Agric J 84:10–12

Nadimpalli BG, Jarret RL, Pathak SC, Kochert G (1992) Phylogenetic relationship of pigeonpea (*Cajanus cajan*) based on nuclear restriction fragment length polymorphisms. Genome 36:216–223

Naithani ST (1941) Cytological studies on Indian pulses. Part I: the somatic chromosomes and the pro chromosomes of *Cajanus*. Proc Natl Acad Sci India Sect B Biol Sci 11:67–73

Nakaya A, Isobe SN (2012) Will genomic selection be a practical method for plant breeding? Ann Bot 110:1303–1316

Odeny DA, Jayashree B, Ferguson M, Hoisington D, Crouch J, Gebhardt C (2007) Development, characterization and utilization of microsatellite markers in pigeonpea. Plant Breed 126:130–136

Odeny DA, Githiri SM, Kimani PM (2009) Inheritance of resistance to Fusarium wilt in pigeonpea (*Cajanus cajan (*L.) Millsp.). J Animal Plant Sci 2:89–95

Ohri D, Singh SP (2002) Karyotypic and genome size variation in *Cajanus cajan* (L.) Millsp. (Pigeonpea) and some wild relatives. Genet Resour Crop Evol 49:1–10

Ongom PO, Adeyanju A, Gobena DJ, Rich P, Ejeta G (2016) Sorghum MAGIC population: structure and potential for genetics research and breeding. Plant and genome conference XXIV, San Diego, CA. January 08–13, 2016, P0767. (https://pag.confex.com/pag/xxiv/webprogram/Paper21680.html)

Pande S, Sharma M, Mangla UN, Ghosh R, Sundaresan G (2011) Phytophthora blight of pigeonpea [*Cajanus cajan* (L.) Millsp.]: an updating review of biology, pathogenicity and disease management. Crop Protect 30:951e957

Patel KA, Acharya S, Prajapati N, Patel JB (2012) Molecular identification of cytoplasmic male sterility based hybrid GTH 1 and its parents in pigeonpea. Indian J Genet 72:94–96

Pazhamala L, Saxena RK, Singh VK, Sameerkumar C et al (2015) Genomics-assisted breeding for boosting crop improvement in pigeon pea (Cajanus cajan). Front Plant Sci 6:50

Priyanka B, Sekhar K, Sunitha T, Reddy VD, Rao KV (2010) Characterization of expressed sequence tags (ESTs) of pigeonpea (*Cajanus cajan* L.) and functional validation of selected genes for abiotic stress tolerance in *Arabidopsis thaliana*. Mol Genet Genomics 283:273–287

Pundir RPS, Singh RB (1986) Inheritance pattern for pod length and ovule number in wide crosses among *Cajanus* and *Atylosia* species. Euphytica 35:649–652

Raju NL, Gnanesh BN, Lekha P, Jayashree B, Pande S, Hiremath PJ et al (2010) The first set of EST resource for gene discovery and marker development in pigeonpea (*Cajanus cajan* L.). BMC Plant Biol 10:45

Ratnaparkhe MB, Gupta VS, VenMurthy MR, Ranjekar PK (1995) Genetic fingerprinting of pigeonpea [*Cajanus cajan* (L.) Millsp.] and its wild relatives using RAPD markers. Theor Appl Genet 91:893–898

Reddy AA (2009) Pulses production technology: status and way forward. Econ Politic Week xliv (52)

Reddy MV, Nene YL (1981) Estimation of yield loss in pigeonpea due to sterility mosaic. In: Proc. Intern. Workshop on Pigeonpea, ICRISAT center, Patancheru, Andra Pradesh, India, pp. 305–312

Reddy MV, Raju TN, Lenne JM (1998) Diseases of pigeonpea. In: Allen DJ, Lenne JM (eds) The pathology of food legumes. CAB International, Cambridge, pp 517–558

Roorkiwal M, Sawargaonkar SL, Chitikineni A et al (2013) Single nucleotide polymorphism genotyping for breeding and genetics applications in chickpea and pigeonpea using the BeadXpress platform. Plant Genome 6:1–10

Roy B (1933) Studies in the development of the female gametophyte in some leguminous crop plants of India. Indian J Agric Sci 3:1098

Sameer Kumar CV, Singh IP, Vijaykumar R, Patil SB, Tathineni R et al (2016) Pigeonpea—A unique jewel in rainfed cropping systems. Legume Perspect 11:8–10

Saxena RK, Prathima C, Saxena KB, Hoisington DA, Singh NK, Varshney RK (2010a) Novel SSR markers for polymorphism detection in pigeonpea (*Cajanus* spp.). Plant Breed 129:142–148

Saxena RK, Saxena K, Varshney RK (2010b) Application of SSR markers for molecular characterization of hybrid parents and purity assessment of ICPH 2438 hybrid of pigeonpea [*Cajanus cajan* (L.) Millspaugh]. Mol Breed 26:371–380

Saxena RK, Penmetsa RV, Upadhyaya HD, Kumar A, Carrasquilla-Garcia N, Schlueter JA et al (2012) Large-scale development of cost-effective single-nucleotide polymorphism marker assays for genetic mapping in pigeonpea and comparative mapping in legumes. DNA Res 19:449–461

Saxena RK, von Wettberg E, Upadhyaya HD, Sanchez V, Songok S, Saxena KB et al (2014) Genetic diversity and demographic history of *Cajanus* spp. illustrated from genome-wide SNPs. PLoS ONE 9:e88568

Saxena RK, Saxena KB, Pazhamala LT, Patel K, Parupalli S, Sameerkumar CV, Varshney RK (2015) Genomics for greater efficiency in pigeonpea hybrid breeding. Front Plant Sci 6:793

Schneeberger K (2014) Using next-generation sequencing to isolate mutant genes from forward genetic screens. Nat Rev Genet 15:662–676

Sharma OP, Bhosle BB, Kamble KR, Bhede BV, Seeras NR (2011) Management of pigeonpea pod borers with special reference to pod fly (*Melanagromyza obtusa*). Indian J Agric Sci 81

Sharma HC, Srivastava CP, Durairaj C, Gowda CLL (2010) Pest management in grain legumes and climate change. In: Yadav SS, McNeil DL, Redden R, Patil SA (eds) Climate change and management of cool season grain legume crops. Springer Science + Business Media, Dordrecht, The Netherlands, pp 115–140

Singh NK, Gupta DK, Jayaswal PK, Mahato AK, Dutta S, Singh S et al (2011) The first draft of the pigeonpea genome sequence. J Plant Biochem Biotechnol 21:98–112

Singh VK, Khan AW, Saxena RK, Kumar V, Kale SM, Sinha P et al (2015) Next-generation sequencing for identification of candidate genes for Fusarium wilt and sterility mosaic disease in pigeonpea (*Cajanus cajan*). Plant Biotechnol J. doi:10.1111/pbi.12470

Singh IP, Bohra A, Singh F (2016a) An overview of varietal development programme of pigeonpea in India. Legume Perspect 11:37–40

Singh D, Sinha B, Rai VP, Singh MN, Singh DK, Kumar R, Singh AK (2016b) Genetics of Fusarium wilt resistance in pigeonpea (*Cajanus cajan*) and efficacy of associated SSR markers. Plant Pathol J 32:95–101

Sinha P, Singh VK, Suryanarayana V, Krishnamurthy L, Saxena RK, Varshney RK (2015a) Evaluation and validation of housekeeping genes as reference for gene expression studies in Pigeonpea (*Cajanus cajan*) under drought stress conditions. PLoS One 10: e0122847

Sinha P, Saxena RK, Singh VK, Varshney RK (2015b) Selection and validation of housekeeping genes as reference for gene expression studies in pigeonpea (*Cajanus cajan*) under heat and salt stress conditions. Front Plant Sci 6:1071

Sinha P, Saxena KB, Saxena RK, Singh VK, Suryanarayana V, Sameer Kumar V et al (2015c) Association of nad7a gene with cytoplasmic male sterility in pigeonpea. Plant Genome 8:1–12

Tayeh N, Aubert G, Pilet-Nayel M, Lejeune-Hénaut I, Warkentin TD, Burstin J (2015) Genomic tools in pea breeding programs: status and perspectives. Front Plant Sci 6:1037

Tester M, Langridge P (2010) Breeding technologies to increase crop production in a changing world. Science 327:818–822

Touzet P, Meyer EH (2014) Cytoplasmic male sterility and mitochondrial metabolism in plants. Mitochondrion 19:166–171

Tuteja R, Saxena RK, Davila J, Shah T, Chen W, Xiao Y et al (2013) Cytoplasmic male sterility associated chimeric open reading frames identified by mitochondrial genome sequencing of four *Cajanus* genotypes. DNA Res 20:485–495

Varshney RK, Penmetsa RV, Dutta S, Kulwal PL, Saxena RK, Datta S et al (2010) Pigeonpea genomics initiative (PGI): an international effort to improve crop productivity 149:113–120

Varshney RK, Chen W, Li Y, Bharti AK, Saxena RK, Schlueter JA et al (2012) Draft genome sequence of pigeonpea (*Cajanus cajan*), an orphan legume crop of resource-poor farmers. Nat Biotechnol 30:83–89

Varshney RK, Murali Mohan S, Gaur PM, Gangarao NVPR, Pandey MK et al (2013) Achievements and prospects of genomics-assisted breeding in three legume crops of the semi- arid tropics. Biotechnol Adv S0734-9750 (13)00003-7

van der Maesen L JG (1980). India is the native home of pigeonpea. In: Arends JC, Boelema G, de Groot CT, Leeuwenberg AJM (Eds) Libergratulatorius in honorem H. C.D. de Wit landbouwhogeschool Miscellaneous paper no 19. Veenman H, Zonen BV, Wageningen, Netherlands, pp 257–262

van der Maesen LJG (1990) Pigeonpea: origin, history, evolution and taxonomy. In: Nene YL, Hall SD, Sheila VK (eds) The Pigeonpea. CAB International, Wallingford, pp 44–86

Vavilov NI (1951) The origin, variation, immunity and breeding of cultivated plants. Chronica Botanica 13:1–366

Zeven AC, Zhukovsky PM (1975) In: Dictionary of cultivated plants and their centres of diversity, Pudoc, Wageningen, the Netherlands, pp 9–26

Molecular Mapping of Genes and QTLs in Pigeonpea

6

Reyazul Rouf Mir, Irshad Ahmad Rather,
Mohd Ashraf Bhat, G.A. Parray and Rajeev K. Varshney

Abstract

Pigeonpea is one of the most important grain legume crops grown in arid and semiarid regions of the world. There is an increasing demand for the development of new cultivars with high yield potential and better adaptability to adverse environmental conditions. Recent advances in genomics tools and techniques have helped to develop large repertoire of molecular markers and genotypic platforms. The availability of molecular markers facilitated the development of high-density genetic maps that have been used in discovery of important/major QTLs for targeted traits in pigeonpea. In addition, the availability of high-throughput genotypic platforms helped to generate whole genome genotypic data in high-throughput manner necessary for whole genome scanning/ genome-wide association mapping of economically important traits. The advances in comparative genomics, transcriptomics, and whole genome sequencing have uncovered thousands of useful genes including some genes unique to pigeonpea crop. The availability of wealth of genomics resources/information will facilitate molecular breeding aimed at improving production and productivity of pigeonpea in extreme environments of arid and semiarid regions of the world.

R.R. Mir (✉) · I.A. Rather · M.A. Bhat
G.A. Parray
Division of Plant Breeding and Genetics,
Sher-e-Kashmir University of Agricultural Sciences
& Technology of Kashmir (SKUAST-K), Kashmir
180009, India
e-mail: imrouf2006@gmail.com

R.K. Varshney
International Crops Research Institute for the
Semi-Arid Tropics (ICRISAT),
Patancheru 502324, India

R.K. Varshney et al. (eds.), *The Pigeonpea Genome*,
Compendium of Plant Genomes, DOI 10.1007/978-3-319-63797-6_6

6.1 Introduction

Pigeonpea (*Cajanus cajan* L. Millsp.) is one of the major grain legume crops of the arid and semiarid regions of the world and the second most important pulse crop in India. It is cultivated in many parts of the world like China, Kenya, Myanmar, Nepal (Smartt 1990), but its primary origin and diversification center is India (Vavilov 1928; Van Der Maesen 1990). Globally, pigeonpea is grown on ~5 million hectares (ha) and India accounts for over 85% of the global area (Saxena et al. 2015). Pigeonpea is a diploid crop with 2n = 2x = 22 and having a genome size of 833.1 Mbp (Varshney et al. 2012a). This crop has great nutritional as well as economic value especially for poor people living in Asia, Africa, South America, Central America, and the Caribbean (Mula and Saxena 2010). Almost all parts of this crop are consumed for variety of purposes like pigeonpea seeds having 20–22% protein are consumed as green peas, whole grain, or split peas, the seed and pod husks make a quality feed, whereas dry branches and stems serve as domestic fuel, fallen leaves from the plant provide vital nutrients to the soil, and the plant also enriches soil through symbiotic nitrogen fixation (Varshney et al. 2010a, 2010b; Saxena et al. 2010a, b; Mohar et al. 2014)

Pigeonpea withstands elevated temperatures and water scarcity and that adds to its importance as a crop of choice in arid and semiarid tropical (SAT) regions of the world. Despite its importance in the SAT regions and extensive efforts of research community, little improvement has been made in increasing pigeonpea production. Its production is mostly hindered by a variety of biotic (diseases and insect pests) and abiotic (drought, salinity, and waterlogging) stresses (Varshney et al. 2010a, b). *Fusarium* wilt (FW) and sterility mosaic disease (SMD) are the major biotic constraints limiting pigeonpea production.

In order to address above production constraints in pigeonpea, genomics tools like molecular markers, genetic/trait mapping, and genome sequencing are the prerequisites for pigeonpea improvement. However, in the past, development of pigeonpea genetic maps has been hindered by low level of genetic diversity present within the gene pool and less availability of DNA-based molecular markers. Little was done in terms of discovery of genes for important targeted traits. This paucity in gene discovery programs in pigeonpea was partly attributed to its 'orphan crop' status few years back. During the last decade, ICRISAT along with its national/international partners has developed significant genomic resources of pigeonpea that is impacting/will impact on trait mapping and molecular breeding of this crop. Therefore, recent advances in genomics tools and technologies like advances in sequencing and genotyping technologies have made this crop as one of the resource-rich legume crops. As a result, large repertoire of molecular markers, genotyping platforms, transcriptomics resources, and more recently draft genome sequence have been developed/became available in this crop (Varshney et al. 2012a, b).

In terms of molecular markers, large sets of simple sequence repeat (SSR) markers (Saxena et al. 2010a, b; Bohra et al. 2011; Dutta et al. 2011), diversity array technology (DArT) markers (Yang et al. 2006, 2011), single feature polymorphism (SFP) markers (Saxena et al. 2011), and single nucleotide polymorphism (SNP) genotyping platforms (see Varshney et al. 2012b; Saxena et al. 2014) have been developed. In addition, molecular markers have been used for the development of low-, moderate-, and high-density genetic maps in pigeonpea. For instance, a moderate-density interspecific genetic map is already available (Bohra et al. 2011) and a less dense genetic map for cultivated pigeonpea has been also developed (Gnanesh et al. 2011). More recently, four genetic maps and subsequent first consensus/integrated genetic map based on six F_2 intraspecific mapping populations were also developed in pigeonpea (Bohra et al. 2012). These molecular markers and molecular marker-based genetic maps that have become available/been developed in pigeonpea provide great opportunities to discover genes/QTLs responsible for important targeted traits in pigeonpea. The discovery of genes once done will facilitate molecular breeding aimed at improving pigeonpea against a range of stresses, quality, etc. Some studies have been already conducted, where these genetic maps have helped in discovery of quantitative trait loci (QTLs)/genes for some of the important traits like sterility mosaic disease (SMD) (Gnanesh et al. 2011) and fertility restoration (FR) (Bohra et al. 2012).

Another very important approach that does not involve development and use of genetic maps from biparental mapping populations for gene discovery is 'association mapping.' Association mapping is considered one of the most important approaches of gene discovery that overcomes several limitations of QTL mapping (for details, see Mir et al. 2012 for a review). This approach has been already used in pigeonpea by scanning whole genome of pigeonpea through Illumina, GoldenGate SNP, and DArT markers, followed by discovery of genes/significantly associated markers linked with determinacy trait in pigeonpea (Mir et al. 2013). Similarly, in another study, whole genome association mapping (using DArT markers) has been also used for identification of marker-trait associations for growth, phonological and yield/yield contributing traits in pigeonpea (Rajeev Varshney, personal communication). In addition to marker-based QTL mapping and association mapping, candidate gene-based association mapping has been also used to identify candidate genes responsible for determinacy trait in pigeonpea. In this chapter, efforts have been made to summarize all the available results on discovery of gene/QTLs through different approaches in pigeon pea. We have also made efforts to summarize results on discovery/identification of genes through transcriptomics/next-generation sequencing approaches in pigeonpea.

6.2 Approaches for Identification and Mapping of Genes and QTLs in Pigeonpea

Several approaches are available and have been used to identify genes and QTLs in pigeonpea (Fig. 6.1). However, the progress of mapping of genes and QTLs in pigeonpea is still at its early stage when compared to other grain legume crops like soybean, chickpea, pea, common bean, and groundnut. Therefore, concerted efforts need to be made to map genes for important agronomic traits in this crop. The genes once mapped could be useful for improving production and productivity of pigeonpea in arid and semiarid regions of the world. For mapping of QTLs, two important approaches like QTL mapping and association mapping have been extensively used in all crops including pigeonpea. The identification of QTLs mainly relied on linkage analysis involving use of biparental mapping populations. This method is more common and proved successful for identification of QTLs for variety of traits in different crops worldwide. However, QTL mapping has several disadvantages including possessing low resolution (see Mir et al. 2012; Gupta et al. 2013). To overcome most of the disadvantages associated with QTL mapping, linkage disequilibrium-based association mapping (AM) has provided alternative strategy to identify marker-trait associations (MTAs). Association mapping has number of advantages over other mapping techniques including the potential for increased QTL resolution, and an increased sampling of molecular variation (for reviews, see Buckler and Thornsberry 2002; Flint-Garcia et al. 2003; Gupta et al. 2005; Yu and Buckler 2006; Mir et al. 2012). In addition to QTL mapping and association mapping for mapping genes and QTLs, more direct methods like transcriptomics, candidate gene analysis/sequencing, and genome sequencing have been preferred method for gene discovery nowadays. The progress made in discovery of genes through various approaches in pigeonpea will be discussed below.

6.3 Identification/Mapping of Genes Through QTL Mapping and Association Mapping

As mentioned earlier, not many studies have been conducted for discovery of QTLs through either QTL mapping or association mapping. Few studies that involve identification of QTLs through QTL mapping/association mapping in pigeon pea are summarized in Table 6.1. The important traits that have been targeted in these few studies included plant height, number of branches, pod number, flowering time, determinacy, fertility restoration (FR), *Fusarium* wilt (FW), and sterility mosaic disease (SMD) (Table 6.1). However, among these traits, most important targeted traits receiving more attention are biotic stresses like FW and SMD. For instance, several RAPD, SCAR, and SSR markers have been reported to be associated with FW in some earlier studies (Kotresh et al. 2006; Prasanthi et al. 2009; Singh et al. 2013). Similarly for SMD, Gnanesh et al. (2011) identified six (06) QTLs in two different populations and for two different SMD isolates. Among these six (06) QTLs, one QTL (qSMD4) on LG07 explaining 24.72% phenotypic variation for SMD is considered major/important QTL. This major QTL may prove useful in marker-assisted selection (MAS) programs after its validation. For FR, four (04) QTLs/genomic regions including two major QTLs (explaining >20% phenotypic variation) were identified from three different

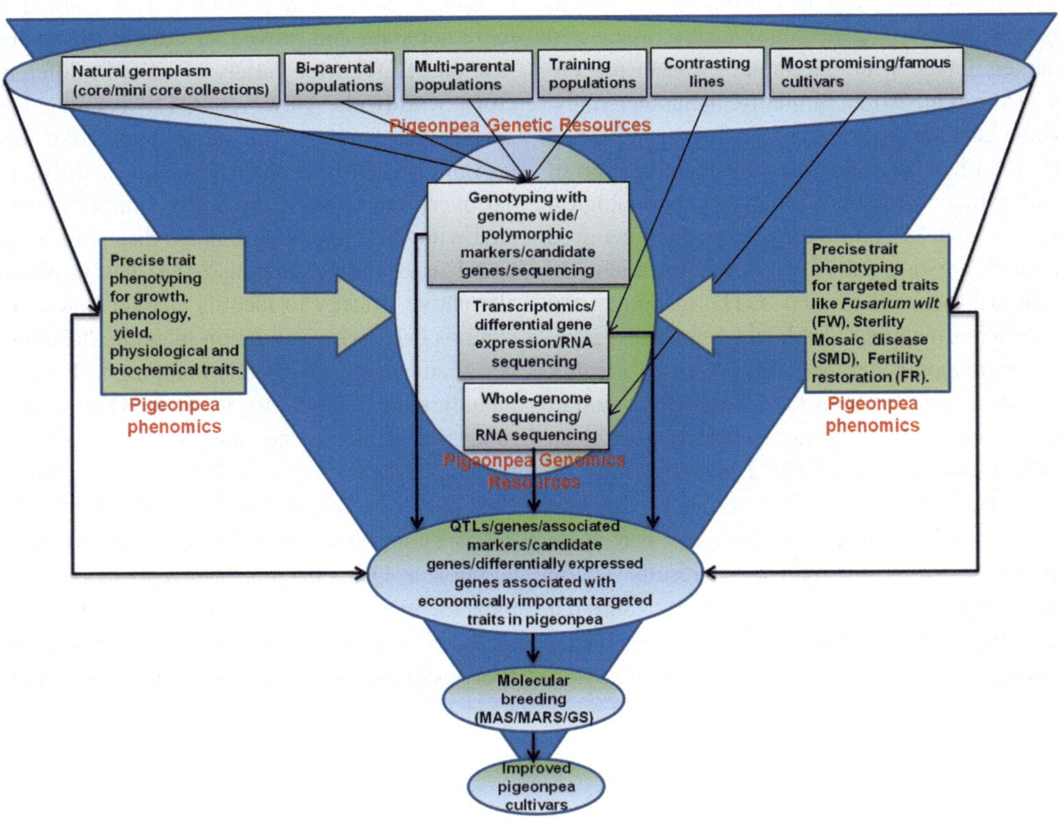

Fig. 6.1 Different steps involved in discovery/mapping of genes/QTLs for different traits and their use for development of improved pigeonpea cultivars

Table 6.1 QTLs identified through QTL mapping and association mapping in pigeon pea

Trait	Name/number of QTL/marker identified	Linkage group (LG)	References
Plant height	***qPH4.1***[a], *qPH5.1*	LG_Cc4, LG_Cc5	Kumawat et al. (2012)
Primary branching	***qPB4.1***[a], *qPB5.1*	LG_Cc4, LG_Cc5	
Secondary branching	*qSB5.1*	LG_Cc5	
Number of pods	*qPD3.1*, *qPD4.1*, ***qPD5.1***[a]	LG_Cc3, LG_Cc4, LG_Cc5	
Flowering	***qFL4.1***[a], *qFL5.1*	LG_Cc4, LG_Cc5	
Maturity	*qMT4.1*, ***qMT5.1***[a], *qMT10.1*	LG_Cc4, LG_Cc5, LG_Cc10,	
Fertility restoration	QTL-RF-1, QTL-RF-2, QTL-RF-3, **QTL-RF-4**[a]	LG06, LG11, LG06, LG06	Bohra et al. (2012)
SMD resistance	qSMD4	LG07	Gnanesh et al. (2011)
Determinacy	Six DArT and 19 SNP markers	NA	Mir et al. (2013)

The QTL with highest R^2 value has been highlighted with 'a'

genetic backgrounds/mapping populations. These QTLs were designated as QTL-RF-1, QTL-RF-2, QTL-RF-3, and QTL-RF-4 explaining phenotypic variation of 14.85, 15.84, 20.89, and 24.17%, respectively (Bohra et al. 2012). For plant height, number of branches, pod number, and flowering time, a total of 13 QTLs were identified (Table 6.1). The phenotypic variation explained (PVE%) by these individual QTLs ranged from 3.18 to 51.4% (Kumawat et al. 2012). Similarly for determinacy, six DArT (among 6144 features) and 19 SNPs (among 768 SNPs) have been found significantly associated with determinacy trait in collection of 94 diverse pigeonpea germplasm lines comprising both determinate (11) and indeterminate (83) lines (Mir et al. 2013).

In addition to QTL mapping and association mapping, sometimes advanced backcross-QTL (AB-QTL) mapping initially proposed by Tanksley and Nelson (1996) has been found promising for simultaneous discovery and transfer of QTLs. In pigeonpea, AB-QTL mapping projects were also started by a team of scientists led by Dr. Rajeev Varshney and Nallini Malikarjuna of ICRISAT, Hyderabad, by developing four different mapping populations involving *C. cajanifolius* and *C. acutifolius* species for agronomically important traits. The mapping populations have been genotyped using DArT markers and are being phenotyped for different traits for their use in identification of MTAs (Rajeev Varshney, personal communication).

6.4 Candidate Gene Analysis/Identification/Mapping in Pigeonpea

Biotic and abiotic stresses cause huge losses in the production of pigeonpea. Among the biotic stresses, FW and SMD are the two major constraints in crop production (Sharma et al. 2012). Several candidate genes have been already identified and mapped for different traits using different plant material/types of populations. The process of candidate gene identification/mapping will be accelerate due to the availability of pigeonpea draft genome sequence assembly. Different approaches like comparative genomics, translational genomics, differential gene expression (qRT-PCR), overexpression, NGS-based approaches, genetic/QTL mapping have been used to identify and map candidate genes in pigeonpea (Table 6.2). For instance, two important genes *C.cajan_01839* and *C.cajan_03203* have been identified for resistance to SMD and FW,

Table 6.2 Candidate genes identified/mapped in pigeonpea

Gene	Trait	Approach	References
CcHyPRP	Drought	Overexpression	Priyanka et al. (2010)
CcCYP	Drought, salinity, and extreme temperatures	Overexpression	Sekhar et al. (2010)
CcMT1	Heavy metal stress	Overexpression	Sekhar et al. (2010)
CcCDR	Drought, salinity, and cold	Homology	Tamirisa et al. (2014)
CcTFL1	Determinacy	Comparative genomics/candidate gene sequencing/linkage mapping	Mir et al. (2014)
DLP	Drought	qRT-PCR	Deeplanaik et al. (2013)

respectively, by using NGS-based approach (Singh et al. 2016). Seven genes such as *CcHyPRP, CcCDR, CcCYP, CcMT1, DLP, APB,* and *LTP1* were identified in pigeonpea from subtracted cDNA libraries among which transcriptional up-regulation of *dehydrin-like protein* (*DLP*) was observed under drought response (Deeplanaik et al. 2013). In another study, seven genes (*CcAP, CcFCA, CcFLD, CcFKF1, CcGI, CcTFL2,* and *CcTFL1*) related to determinacy/flowering pattern in pigeonpea were isolated through a comparative genomics approach. Among these seven genes, two genes (*CcGI* and *CcTFL1*) were genetically mapped using biparental mapping population and one gene (*CcTFL1*) was declared likely candidate for determinacy in pigeonpea by using candidate gene sequencing, comparative genomics, linkage-based QTL mapping, and expression analysis approaches (Mir et al. 2014).

6.5 Identification of Genes Through Transcriptomics

The traditional molecular mapping methods have been extensively used to map quantitative trait loci (QTL)/genes for traits of agronomical importance. However there use have been limited, due to time consumption, expensive and laborious procedures for marker development and validation. This resulted in advancement toward efficient, accurate, and cost-effective strategy of next-generation sequencing (NGS) technologies. NGS technologies have opened up wealth of opportunities for plant breeding and genetics studies in addition to de novo sequencing of crops. Transcriptome sequencing has increased the accessibility of genomic resources in pigeonpea. In pigeonpea, the first EST resource provides the transcriptomics for gene discovery and development of functional markers associated with *Fusarium* wilt and SMD (Raju et al. 2010). As root is the site of infection for *Fusarium udum*, therefore, cDNA libraries were constructed from root tissues to evaluate the transcriptional response after pathogen infection. Similarly for SMD, leaf is the primary site of infection and as such cDNA libraries were constructed from leaf tissues. A total of four pigeonpea genotypes namely 'ICPL 20102' (FW resistant), 'ICP 2376' (FW susceptible), 'ICP7035' (SMD-resistant), and 'TTB7' (SMD-susceptible) were used in this study. Another set of ESTs was generated from whole plant tissues of pigeonpea subjected to PEG/water deficit stress conditions (Priyanka et al. 2010). Two subtracted cDNA libraries were constructed from pigeonpea plants treated with 10% polyethylene glycol (PEG-6000). Among the various ESTs identified, three of the selected stress-responsive genes, viz. *CcHyPRP*, *CcCDR*, and *CcCYP* showed remarkable tolerance against multiple abiotic stresses in transgenic *Arabidopsis*. Kumar et al. (2015) analyzed the transcriptome of pigeonpea roots under

Table 6.3 NGS platforms used for identification of transcripts/genes in pigeonpea

NGS platform	Transcripts/unigenes	References
454FLX	43,324	Dutta et al. (2011)
454FLX and Illumina	127,754	Dubey et al. (2011)
Illumina	–	Saxena et al. (2012)
FLX/454 and Illumina	21,434	Kudapa et al. (2012)

water deficit by suppression subtractive hybridization (SSH). In this experiment, plants were grown at –0.45 MPa stress level by using PEG-6000 for SSH library construction. A comprehensive transcriptome assembly from more than 16 pigeonpea genotypes has been developed by using Sanger along with second-generation sequencing platforms (Illumina GA IIx and FLX/454) (Kudapa et al. 2012). The resultant transcriptome comprised of 21,434 transcript assembly contigs. This transcriptome referred as *Cc*TAv2 was generated from a range of tissues and under various stress treatments given to the plants. *Cc*TAv2 was mapped on the soybean genome as reference, and putative mapping positions of ISR markers (intron spanning regions) were predicted. A subset of ISR markers were validated in few parental genotypes for important economic traits that are important for pigeonpea improvement. *Cc*TAv2 assembly can be used for gene identification and function, for development of molecular markers like SSRs, SNPs (Varshney et al. 2009). In general, transcriptome sequencing has increased the genomic resource availability in pigeonpea, which can be used for molecular breeding programs to develop elite cultivars of pigeonpea. Different sequencing platforms used to generate transcriptome assemblies in pigeonpea are summarized in Table 6.3.

6.6 Genes Identified/Mapped Through Genome Sequencing

Low genetic diversity among the cultivated pigeonpea genotypes has hindered the conventional breeding programs in this crop. Before 2006, almost no genomics resource of pigeonpea was available. Pigeonpea genomics initiative (PGI) under the umbrella of Indo-US agricultural knowledge initiative (AKI) was started in late 2006 to enable genomics-assisted breeding in this crop. Later funding from the Generation Challenge Program (GCP) of the Consultative Group on International Agricultural Research (CGIAR) and by joining of additional collaborators across several institutes and countries has strengthened the initiative. Varshney et al. (2012a) used the Illumina GA and HiSeq 2000 sequencing platforms along with Sanger-based bacterial artificial chromosome (BAC) end sequencing to decode 605.78 Mb (72.7%) of the 833.07 Mb pigeonpea genome. This represents the first genomic report of orphan legume crop 'pigeonpea' with 48,680 genes. Of the 48,680 genes, 266 were identified as ORFans (genes representing lineage specific pattern and with no sequence similarity in the genome or protein database) and 111 drought-responsive genes.

Another draft genome of the pigeonpea variety, 'Asha,' was completed by using 454 GS-FLX chemistry (Singh et al. 2012). This study identified 59,515 genes including 12, 511 transposable element-related genes. Among 47,004 protein-coding genes, 1213 were identified as disease resistance/defense response and 152 as abiotic stress tolerance genes. Availability of genome sequence has opened a wealth of opportunities for the improvement of pigeonpea. In pigeonpea, next-generation sequencing, sequencing-based bulked segregant analysis (Seq-BSA) was used to map resistance genes for FW and SMD. Two important genes namely *C.cajan*_01839 for SMD resistance and *C.cajan*_03203 for FW resistance have been reported in pigeonpea, whose introgression will be useful in molecular breeding of the crop (Singh et al. 2016).

6.7 Conclusion

Gradual decline in soil water and increase in temperature will be the major challenges in near future for food production. Drought and high-temperature tolerant crops such as pigeonpea can be a choice to provide food security especially to the SAT regions of the world. Low genetic diversity and biotic stresses are the major bottleneck in pigeonpea production. However, significant progress has been made in the generation of vast genomics resources, transcriptomics, and whole genome sequencing. In addition, the availability of wild/cultivated germplasm, core/mini-core collections, biparental mapping populations, advanced breeding lines, released varieties has facilitated discovery of genes/QTLs in pigeonpea. Various approaches of gene discovery like QTL mapping, association mapping, candidate gene sequencing/analysis, transcriptomics, and whole genome sequencing have been employed successfully in this crop (Fig. 6.1). With the result promising genes/QTLs with major effect have become available for important targeted traits like flowering time, determinacy, fertility restoration (FR), *Fusarium* wilt (FW), and sterility mosaic disease (SMD). More sophisticated approaches like use of nested association mapping (NAM) and use of multi-parent advanced generation intercross populations (MAGIC) shall be used in future for high resolution/precision mapping of genes/QTLs for important traits in pigeonpea. The genes/QTLs already identified shall be validated in different genetic backgrounds before recommending them for marker-assisted selection (MAS) programs in pigeonpea. In addition to marker-assisted selection (MAS), other modern breeding approaches like marker-assisted recurrent selection (MARS) and genomic selection (GS) shall be used in pigeonpea for quick genetic gain.

References

Bohra A, Dubey A, Saxena RK, Penmetsa RV, Poornima KN, Kumar N, Farmer AD, Srivani G, Upadhyaya HD, Gothalwal R, Ramesh R et al (2011) Analysis of BAC-end sequences (BESs) and development of BES-SSR markers for genetic mapping and hybrid purity assessment in pigeon pea (*Cajanus* spp.). BMC Plant Biol 11:56

Bohra A, Saxena RK, Gnanesh BN, Saxena KB, Byregowda M, Rathore A et al (2012) An intra-specific consensus genetic map of pigeon pea [*Cajanus cajan* (L.) Millspaugh] derived from six mapping populations. Theor Appl Genet 125:1325–1338

Buckler ES, Thornsberry JM (2002) Plant molecular diversity and applications to genomics. Curr Opin Plant Biol 5:107–111

Deeplanaik N, Kumaran RC, Venkatarangaiah K, Shivashankar SKH, Doddamani D, Telkar S (2013) Expression of drought responsive genes in pigeon pea and in silico comparison with soybean cDNA library. J Crop Sci Biotechnol 16:243–251

Dubey A, Farmer A, Schlueter J, Cannon SB, Abernathy B, Tuteja R et al (2011) Defining the transcriptome assembly and its use for genome dynamics and transcriptome profiling studies in pigeon pea (*Cajanus cajan* L.). DNA Res 18:153–164

Dutta S, Kumawat G, Singh BP, Gupta DK, Singh S, Dogra V, Gaikwad K, Sharma TR, Raje RS, Bandhopadhya TK, Datta S, Singh MN et al (2011) Development of genic-SSR markers by deep transcriptome sequencing in pigeon pea [*Cajanus cajan* (L.) Millspaugh]. BMC Plant Biol 11:17

Flint-Garcia SA, Thornsberry JM, Buckler ES (2003) Structure of linkage disequilibrium in plants. Annu Rev Plant Biol 54:357–374

Gnanesh BN, Bohra A, Sharma M, Byregowda M, Pande S, Wesley V et al (2011) Genetic mapping and quantitative trait locus analysis of resistance to sterility mosaic disease in pigeon pea [*Cajanus cajan* (L.) Millsp.]. Field Crops Res 123:53–61

Gupta PK, Rustgi S, Kulwal PL (2005) Linkage disequilibrium and association studies in higher plants: present status and future prospects. Plant Mol Biol 57:461–485

Gupta PK, Kulwal PL, Mir RR (2013) QTL mapping: methodology and applications in cereal breeding. In: Gupta PK, Varshney RK (eds) Cereal Genomics II. Springer, Netherlands, pp 275–318

Kotresh H, Fakrudin B, Punnuri SM, Rajkumar BK, Thudi M, Paramesh H et al (2006) Identification of two RAPD markers genetically linked to a recessive allele of a Fusarium wilt resistance gene in pigeon pea (*Cajanus cajan* L. Millsp.). Euphytica 149:113–120

Kudapa H, Bharti AK, Cannon SB, Farmer AD, Mulaosmanovic B, Kramer R et al (2012) A comprehensive transcriptome assembly of pigeon pea (*Cajanus cajan* L.) using Sanger and second-generation sequencing platforms. Mol Plant 5:1020–1028

Kumar R, Yadav S, Shrinivas, Kumar Srivastava A, Shitole V, Naik GR (2015) Transcriptome of pigeon pea roots under water deficit analyzed by suppression subtractive hybridization. J Agri Sci Technol 17:1333–1345

Kumawat G, Raje RS, Bhutani S, Pal JK, Mithra ASVCR, Gaikwad K, Sharma TR, Singh NK (2012) Molecular mapping of QTLs for plant type and earliness traits in pigeon pea (*Cajanus cajan* L. Millsp.). BMC Genet 13:84

Mir RR, Zaman-Allah M, Sreenivasulu N, Trethowan R, Varshney RK (2012) Integrated genomics, physiology and breeding approaches for improving drought tolerance in crops. Theor Appl Genet 125:625–645

Mir RR, Saxena RK, Saxena KB, Upadhyaya HD, Kilian A, Cook DR et al (2013) Whole-genome scanning for mapping determinacy in pigeon pea (*Cajanus* spp.). Plant Breed 132:472–478

Mir RR, Kudapa H, Srikanth S, Saxena RK, Sharma A, Azam S et al (2014) Candidate gene analysis for determinacy in pigeon pea (*Cajanus* spp.). Theor Appl Genet 127:2663–2678

Mohar S, Narinder KG, Mukesh KR, Om PD, Dutta M et al (2014) Pigeon pea genetic resources and its utilization in India:current status and future prospects. J Plant Sci Res 1:107

Mula MG, Saxena KB (2010) Lifting the level of awareness on pigeon pea-a global perspective. International Crops Research Institute for the Semi-Arid Tropics (ICRISAT), Patancheru, India

Prasanthi L, Reddy BVB, Rekha Rani K, Naidu PH (2009) Molecular marker for screening *Fusarium* wilt resistance in pigeon pea [*Cajanus cajan* (L.) Millspaugh]. Legume Res 32:19–24

Priyanka B, Sekhar K, Sunita T, Reddy VD, Rao KV (2010) Characterization of expressed sequence tags (ESTs) of pigeon pea (*Cajanus cajan* L.) and functional validation of selected genes for abiotic stress tolerance in *Arabidopsis thaliana*. Mol Genet Genomics 283:273–287

Raju NL, Gnanesh BN, Lekha P, Jayashree B, Pande S, Hiremath PJ et al (2010) The first set of EST resource for gene discovery and marker development in pigeon pea (*Cajanus cajan* L.). BMC Plant Biol 10:1

Saxena KB, Ravikoti VJ, Sultana R (2010a) Quality nutrition through pigeon pea—a review. Health 2:1335–1344

Saxena RK, Prathima C, Saxena KB, Hoisington DA, Singh NK et al (2010b) Novel SSR markers for polymorphism detection in pigeon pea (*Cajanus spp.*). Plant Breed 129:142–148

Saxena RK, Cui X, Thakur V, Walter B, Close TJ, Varshney RK (2011) Single feature polymorphisms (SFPs) for drought tolerance in pigeon pea (*Cajanus* spp.). Funct Integr Genomics 11:651–657

Saxena RK, Penmetsa RV, Upadhyaya HD, Kumar A, Carrasquilla-Garcia N, Schlueter J et al (2012) Large-scale development of cost-effective single-nucleotide polymorphism marker assays for genetic mapping in pigeon pea and comparative mapping in legumes. DNA Res 19:449–461

Saxena RK, von Wettberg E, Upadhyaya HD, Sanchez V, Songok S, Saxena KB et al (2014) Genetic diversity and demographic history of *Cajanus* spp. illustrated from genome-wide SNPs. PLoS ONE 9:e88568

Saxena RK, Saxena KB, Pazhamala LT, Patel K, Parupalli S, Sameerkumar CV, Varshney RK (2015) Genomics for greater efficiency in pigeon pea hybrid breeding. Front Plant Sci 6:793

Sekhar K, Priyanka B, Reddy VD, Rao KV (2010) Isolation and characterization of a pigeon pea cyclophilin (CcCYP) gene, and its over-expression in Arabidopsis confers multiple abiotic stress tolerance. Plant, Cell Environ 33:1324–1338

Sharma M, Rathore A, Mangala UN, Ghosh R, Sharma S, Upadhyay HD et al (2012) New sources of resistance to *Fusarium* wilt and sterility mosaic disease in a mini-core collection of pigeon pea germplasm. Eur J Plant Pathol 133:707–714

Singh NK, Gupta DK, Jayaswal PK, Mahato AK, Dutta S, Singh S et al (2012) The first draft of the pigeon pea genome sequence. J Plant Biochem Biotechnol 21:98–112

Singh AK, Rai VP, Chand R, Singh RP, Singh MN (2013) Genetic diversity studies and identification of SSR markers associated with *Fusarium* wilt (*Fusarium udum*) resistance in cultivated pigeon pea (*Cajanus cajan*). J Genet 92:273–280

Singh VK, Khan AW, Saxena RK, Kumar V, Kale SM, Sinha P, et al. (2016) Next-generation sequencing for identification of candidate genes for *Fusarium* wilt and sterility mosaic disease in pigeon pea (*Cajanus cajan*). Plant Biotechnol J 14:1183–1194

Smartt J (1990) Grain legumes: evaluation and genetic resources. Cambridge University Press, Cambridge, UK

Tamirisa S, Vudem DR, Khareedu VR (2014) Overexpression of pigeon pea stress-induced cold and drought regulatory gene (*CcCDR*) confers drought, salt, and cold tolerance in Arabidopsis. J Exp Bot 65:4769–4781

Tanksley SD, Nelson JC (1996) Advanced backcross QTL analysis: a method for the simultaneous discovery and transfer of valuable QTLs from unadapted germplasm into elite breeding lines. Theor Appl Genet 92:191–203

Van Der Maesen LJG (1990) Pigeon pea: Origin, history, evolution and taxonomy. In: Nene YL, Hall SD, Sheila VK (eds) The Pigeon pea. CAB International, Wallingford, UK, pp 15–46

Varshney RK, Nayak SN, May GD, Jackson SA (2009) Next generation sequencing technologies and their implications for crop genetics and breeding. Trends Biotechnol 27:522–530

Varshney RK, Penmetsa RV, Dutta S, Kulwal PL, Saxena RK, Datta S et al (2010a) pigeon pea genomics initiative (PGI): an international effort to improve crop productivity of pigeon pea (*Cajanus cajan* L.). Mol Breed 26:393–408

Varshney RK, Thundi M, May GD, Jackson SA (2010b) Legume genomics and breeding. Plant Breed Rev 33:257–304

Varshney RK, Chen W, Li Y, Bharti AK, Saxena RK et al (2012a) Draft genome sequence of Pigeon pea (*Cajanus cajan*), an orphan legume crop of resource-poor farmer. Nat Biotechnol 30:83–89

Varshney RK, Kudapa H, Roorkiwal M, Thudi M, Pandey MK, Saxena RK et al (2012b) Advances in genomics research and molecular breeding applications in SAT legume crops by using next generation sequencing and high-throughput genotyping technologies. J Biosci 37:811–820

Vavilov (1928) Geographische Genzentren unserer Kulturpflanzen; Verhandlungen des V Internationalen Kongresses fur Vererbungswissenschaft. Berlin, Germany 1927: 342–369

Yang S, Pang W, Harper J, Carling J, Wenzl P, Huttner E et al (2006) Low level of genetic diversity in cultivated pigeon pea compared to its wild relatives is revealed by diversity arrays technology (DArT). Theor Appl Genet 113:585–595

Yang S, Saxena RK, Kulwal PL, Ash GJ, Dubey A, Harper JD et al (2011) First genetic map of pigeon pea based on diversity array technology (DArT) markers. J Genet 90:103–109

Yu J, Buckler ES (2006) Genetic association mapping and genome organization of maize. Curr Opin Biotechnol 17:155–160

Germplasm Characterization and Trait Discovery

7

Christopher P. Krieg, Mulualem T. Kassa and Eric J.B. von Wettberg

Abstract

Diverse germplasm is essential to breeding new pigeonpea varieties resilient against both abiotic and biotic challenges. In this chapter, we review the major abiotic and biotic challenges faced by pigeonpea, and trait assessment tools available to the breeding community to address these challenges. We place particular attention on drought tolerance, due to its widespread nature and large effect on yield. We emphasize the utility of wild material to expand the range of available genetic variation; as such material is generally underutilized in breeding programs. A complete genome brings great potential to the pigeonpea breeding community; to fulfill the promise of a genome for breeding for climate resilience, the full range of diversity available must be brought to bear. We hope that conditions in the future allow wild germplasm collections to expand, further capitalizing on the potential to increase pigeonpea yields and resilience against climate change.

7.1 Introduction

Since the advent of agriculture over 10,000 years ago, the domestication of plants by artificial selection and directed breeding has dramatically increased crop yields and altered plant pheno-

C.P. Krieg · E.J.B. von Wettberg (✉)
Department of Biological Sciences and International Center for Tropical Botany, Florida International University, 33199, Miami, FL, USA
e-mail: ebishopv@uvm.edu

C.P. Krieg
Department of Botany, University of Florida, Gainesville, FL, USA

M.T. Kassa
National Research Council, 110 Gymnasium Place, Saskatoon, SK S7N 0W9, Canada

M.T. Kassa
Agriculture and Agri-Food Canada, Morden Research and Development Centre, 101 Route 100, Morden, MB R6M 1Y5, Canada

E.J.B. von Wettberg
Department of Plant and Soil Science, University of Vermont, 63 Carrigan Drive, Burlington, VT 05405, USA

R.K. Varshney et al. (eds.), *The Pigeonpea Genome*,
Compendium of Plant Genomes, DOI 10.1007/978-3-319-63797-6_7

types such as flowering time, growth form and size, biotic stress responses (e.g., herbivore and disease tolerance), and abiotic stress responses (e.g., heat, drought, salinity tolerance) (Tang et al. 2010; Brown et al. 2009; Purugganan and Fuller 2009; Vaughan et al. 2007). However, the selection of favorable alleles from a population (i.e., artificial selection) can have additional consequences by reducing genetic diversity, reducing adaptive capacity, and maintaining the presence of deleterious alleles (Olsen and Wendel 2013). Crop cultivars, therefore, face genetic constraints that increase vulnerability to variable environments including disease, pests, and abiotic conditions, which impose obstacles to breeding programs tasked to feed a growing population.

To combat the challenges of domestication and substantial reductions of genetic variation within crop lines, the use of diverse germplasm including landraces and wild relatives of crops has proven crucial to crop breeding programs by making additional genetic material available for introgression into crop cultivars and has led to considerable improvement of disease and pest resistance for more than half a century (Vavilov 1922; Hajjar and Hodgkin 2007). While some crop species have substantial germplasm collections of wild relatives, only 2–6% of all international germplasm collections are from crop wild relatives (Maxted and Kell 2009). The lack of sufficient genetic resources poses a significant challenge to breeders and agricultural scientists who are tasked to not only increase crop productivity to feed a growing global population but also accomplish this task in the face of a changing global climate that threatens to exacerbate increasing rates of famine, nutrition related health issues, and threats to conservation of biodiversity. The genetic reservoirs that wild relatives offer are under threat of extinction from habitat fragmentation, and habitat loss (Ford-Lloyd et al. 2011). Climate change further threatens all plant populations and makes abiotic stress tolerance a major target for crop improvement.

Many of the most globally important crops have benefitted from collections of wild relative germplasm. Yet many legumes, particularly those primarily grown in tropical regions, have received less attention. One of these tropical legumes, pigeonpea, *Cajanus cajan*, is the only cultivated member of the genus (Kassa et al. 2016), while the remaining wild relatives belong to the secondary or tertiary levels based on the gene pool concept of Harlan and De wet (1971). In pigeonpea, domestication is somewhat incomplete, as indehiscent seeds remain a breeding challenge. With a better understanding of evolutionary relationships in *Cajanus* and new wild relative collections, strategies can be developed to systematically sample the variation in wild relatives and utilize that variation in breeding programs (Warschefsky et al. 2014). For example, species in the tertiary pool of *Cajanus* have a wide eco-geographical range in Australia and remain an untapped reserve of adaptive potential (Varshney et al. 2011). Moreover, pigeonpea is locally adapted to a wide range of edaphic and climatic conditions around the world, presenting diverse adaptations that could be incorporated into germplasm collections. In this chapter, we focus specifically on pigeonpea (*Cajanus cajan*) and promote an integrative approach to develop agroecology programs capable of feeding an exploding human population in a changing global climate, by (1) considering evolutionary insight and population genetic theory to guide germplasm collection, (2) using physiological tools to understand plant responses to biotic and abiotic stresses, and (3) employ modern molecular tools for phenotyping and trait selection. We intend to review current literature and promote the integration of cross-disciplinary approaches from physiology, population genetics, ecology, and genomics.

7.1.1 Pigeonpea and Its Wild Relatives

Pigeonpea is a short-lived perennial shrub with relatively high levels of abiotic and biotic stress resistance. Pigeonpea is a short-lived perennial belonging to the Leguminosae genus *Cajanus* which is composed of 34 species. Pigeonpea is the only cultivated member of the genus, while the remaining wild relatives were placed in the secondary or tertiary gene pools according to the gene pool concept of Harlan and de Wet (1971). Species in the secondary gene pool comprises the putative progenitor of Pigeonpea, *Cajanus cajanifolius* and others such as *Cajanus albicans*, *C. lineatus*, *C. scarabaeoides*, *C. sericeus*, *C. acutifolius*, *C. confertiflorum*, *C. lanceolatus*, *C. reticulatus,* and *C. trinervius* (Mallikarjuna et al. 2011). Species assigned to the tertiary gene pool includes *Cajanus platycarpus*, *C. mollis,* and *C. crassus* (Upadhyaya et al. 2013).

In terms of geographical distribution, most of the species of the genus *Cajanus* are endemic to either southern/southeastern Asia or northern Australia (Fortunato 2000). Among these, 16 *Cajanus* species occur in Asia (8 of which are endemic to India), 15 species in Australia (of which 13 are endemic), one species of *Cajanus* is confined to West Africa, and 2 species (including *Cajanus cajan*) are ubiquitous throughout the old world.

Using morphological and ecological characters such as habit, leaf structure, hairiness, pod size, strophiole characters and other traits, van der Maesen (1986) grouped the genus *Cajanus* into six sections vis-à-vis Cajanus (2 species), Atylosia (7 species), Fruticosa (9 species), Cantharospermum (5 species), Volubilis (6 species), and Rhynchosoides (3 species). Species in sections Cajanus, Atylosia, and Fruticosa have erect growth habit, Cantharospermum, Volubilis, and Rhynchosoides are climbing and creeping species, and Rhynchosoides are trailing species. Three *Cajanus* species have been further subdivided into botanical varieties: *C. scarabaeoides* into var. *pedunculatus* and var. *scarabaeoides*, *C. reticulatus* into var. *grandifolius*, var. *reticulatus*, and var. *maritimus*, and *C. volubilis* into var. *burmanicus* and var. *volubilis* (van der Maesen 1986). Molecular phylogeny of the genus *Cajanus* resolved three distinct clades: Indian, Australian, and *Scarabaeoides* (Kassa et al. 2012).

Pigeonpea is a hardy, widely adapted and drought tolerant pulse crop cultivated primarily by subsistence farmers on 5.32 million hectares of land (FAO 2012) of semiarid tropics and subtropics in south Asia (mainly on the Indian subcontinent), Africa, and Latin America. India accounts 72.5% of the global production area. Pigeonpea is a highly adaptable grain legume that can promote food security in rain-fed agriculture because it tolerates drought, requires very minimal inputs to give a sustainable yield and is resilient to very harsh biotic and abiotic stresses. In fact, a recent study has identified pigeonpea to be more heat and drought tolerant than the majority of grain legume crops (Khoury et al. 2015). As a legume crop, it plays a major role in fixing atmospheric nitrogen through symbiotic nitrogen fixation with soil bacteria as well as through solubilizing of soil-bound phosphorus and thus improves the nutrient status of the soil (Saxena 2008). In addition to its main use as de-hulled split peas ("dhal") which is the primary source of dietary protein (20–24% per seed) for millions of resource-poor people around the world, pigeonpea also has other uses: its immature seeds and pods are consumed fresh as green vegetables, and stems are used as domestic fuel wood and for making huts and leaves are used as quality fodder (Saxena et al. 2006). In spite of having typical Papilionoideae flowers with a "banner and keel" shape, pigeonpea exhibits a considerable variation (20–70%) in natural insect-aided out-crossing rate and is a partially cross-pollinated species (Saxena et al. 1990). This considerable out-crossing rate may have two major impacts on pigeonpea agronomy and breeding. It creates a problem for maintaining genetic purity in cultivar development, but on the other hand, it has been used effectively in developing elite hybrid varieties through hybridization (Saxena 2008). Some studies have reported that insect species also regularly visited the wild relatives of pigeonpea and noted a few naturally out-crossed plants with distinct traits (Saxena and Kumar 2010). Pigeonpea exhibits high levels of phenotypic and morphological diversity in terms of vegetative, floral, with a wide difference in days to maturity (90–300 days), photoperiod sensitivity, growth habit, and other phenotypic and agronomic traits

(Upadhyaya et al. 2007; Saxena 2008). Pigeonpea germplasm comprises diverse sets of landraces and heterogeneous feral forms with extensive morphological diversity. There are determinate, semi-determinate, and indeterminate genotypes that are adapted to various agroecological settings (Upadhyaya et al. 2007). Regardless of this extensive morphological and phenotypic diversity, molecular genetic analyses in pigeonpea, and wild relatives revealed that there is very low genetic diversity in the domesticated gene pool as compared to the wild groups (Kassa et al. 2012 and references therein; Saxena et al. 2014a, b). This strikingly low polymorphism within the domesticated accessions (including the landraces) signals the severity of the "genetic bottleneck" which happened during pigeonpea domestication. To broaden the genetic base of this highly constrained and narrowed genetic diversity in the domesticated gene pool, there is a need to utilize the high genetic diversity that present in the wild gene pool. The wild relatives are a potential source of novel alleles that can be exploited in breeding and improvement programs in pigeonpea (Saxena et al. 2014b).

7.2 Genetic and Genomic Resources of Pigeonpea and Its Wild Relatives

Pigeonpea accessions have been collected and deposited at various gene banks. The International Crop Research Institute for the Semi-Arid Tropics (ICRISAT) holds 13,771 pigeonpea accessions (Gowda et al. 2013). About 11,221 accessions are deposited at the India National Bureau of Plant Genetic Resources (NBPGR) (Singh et al. 2014). Pigeonpea accessions are also deposited elsewhere including 4116 accessions at US Department of Agriculture (USDA) and 1288 accessions at National Genebank of Kenya (Singh et al. 2013). ICRISAT also holds 555 accessions of wild relatives represented by 67 species belonging to the secondary and tertiary gene pool (Upadhyaya et al. 2011). Various cultivated and wild accessions have been characterized for important agronomic traits including yield, early maturity, high protein content, tolerance to salinity, and drought (Gowda et al. 2013). However, recent studies have shown that the bulk of pigeonpea wild relatives are broadly underrepresented in gene banks and other ex-situ conservation settings and thus are high priority species for further collection to tap important traits to pigeonpea breeders (Khoury et al. 2015).

In the past decade, tremendous progress had been made to enrich genetic and genomic resources of pigeonpea, including the completion of the pigeonpea genome (Varshney et al. 2012). This enables discovery of genes and quantitative trait loci (QTLs) associated with important agronomic traits and development of molecular markers that promote marker assisted breeding. About 30 biparental mapping populations segregating for abiotic and biotic traits have been developed (Varshney et al. 2010). These remain a fantastic resource for trait determination and accelerated molecular breeding. Moreover, efforts are underway to develop multi-parent mapping populations including multi-parent advanced generation intercross (MAGIC) and nested association mapping (NAM) populations (Pazhamala et al. 2015).

Similar effort is needed to expand the range of germplasm resources. With primary through tertiary gene pool species poorly collected in centers of diversity in south Asia and Australia, there is room to greatly expand collections. We advocate for efforts to collect the large aridity and soil type gradients that characterize both south Asia and Western Australia. In both south Asia and Australia, hierarchical sampling across regions has great potential to capture adaptations to drought and other climatic factors that segregate on large spatial scales of degrees of latitude, as well as biotic adaptations such as pest and disease adaptation that likely segregate within populations of wild species. We believe such efforts, if integrated into pre-breeding programs guided by genomic information, have great potential to expand climatic resilience and disease resistance in pigeonpea (e.g., Warschefsky et al. 2014).

7.2.1 Leading Breeding Challenges: Drought, Osmotic Stress, and Other Abiotic Challenges

Of particular importance in a changing climate is understanding plant physiological responses to variable and harsh environments. The relationship between water use, photosynthesis, and plant growth has been studied for over a century by both ecologists and agronomists and offers critical tools for developing agroecology programs. Although the goals of ecologists and agronomists may differ, the underlying physiological theory serves as a common core of principles, equations, and processes that when understood can be applied to a diverse set of fields. It is critical for agronomists to understand the physiological principles that describe how water availability affects plant carbon acquisition, growth, and resource allocation (reviewed in Chaves et al. 2003; Farooq et al. 2009) because plant physiology offers useful tools to assess the health of crop plants, and manipulate agroecological breeding programs to increase yield and plant productivity. The following section of this chapter will review the core physiological principles at the intersection of plant water relations, gas exchange, and productivity that inform agroecology programs and crop breeders of the utility of variation in physiological tolerance and traits. Specific attention is paid to pigeonpea, and we emphasize that understanding the variation in physiological and morphological phenotypes in crop wild relatives can be used to inform germplasm collection efforts.

Harsh environments can cause significant yield losses and negatively impact biomass, pod number, seed yield, seed weight, and quality in cultivated pigeonpea (Toker et al. 2007; Khan et al. 2010; Toker and Mutlu 2011; Hasanuzzaman et al. 2013; Pagano 2014). One of the most pressing issues for global food security is water availability because the agriculture industry accounts for some 70% of total freshwater consumption (FAO 2012), and the irrigation water sources and rain patterns that feed agricultural land are vulnerable to the effects of climate change. Already, 60% of all crop production suffers from drought conditions (Grant 2012). In many developing regions of the world, water is a limiting resource and therefore plant water use and traits related to abiotic stress tolerance have been a significant focus of agricultural scientists and agronomists concerned with maintaining productivity in harsh conditions, including water deficit conditions. In response to the numerous climate models that predict changes in global precipitation and increasing drought severity and durations, a firm understanding of plant water relations is critical for improving crop productivity in water-limited environments (see Ehlers and Goss 2003).

The primary way water leaves a plant is through stomata via transpiration. Transpiration generates a pulling force on the water column in xylem generating negative absolute pressures that can surpass a perfect vacuum. This is possible because water is a polar molecule that forms hydrogen bonds between positive and negative hydrogen atoms that allow molecules to withstand tension. This is known as the cohesion–tension theory and describes the force between water molecules that allow water columns to remain intact when "pulled" through the xylem. When water is limited, plants generate more negative tension within their xylem and are more vulnerable to cavitation and embolism. Cavitation is the breaking of a water column at very negative water potentials to pull gas out of solution and introduces gaseous bubbles into the xylem. An embolism is the formation and spreading of air bubbles and imposes an additional resistance to hydraulic flow and can effectively stop water movement causing wilting and premature senescence. Thus, selective breeding on traits that target efficient water use and drought tolerance are of great importance in all crop systems.

The transpiration of water vapor through stomata is influenced by light environment, heat, and leaf characteristics like hairs and thick epicuticular wax that limit water loss (Holmes and Keiller 2002) offering leaf level traits for plant physiologists and breeders to get more "crop per drop" (Yoo et al. 2009). Guided collection efforts and artificial selection on traits that confer more efficient water use are beneficial in dry or otherwise water-limited conditions, and there are significant tools available to assist in these efforts. Carbon isotope discrimination has long been established as a tool to evaluate plant water-use strategies (Farquhar et al. 1989). In particular, analysis of ^{13}C discrimination is an accurate way to measure water-use efficiency and has been shown to be heritable in several species (Schuster et al. 1992; Donovan and Ehleringer 1994). Carbon isotope discrimination is a superior method to assess water-use efficiency in plants because it is an integrated measure over the life of the sample, and thus ^{13}C discrimination is called often as integrated water-use efficiency (WUE). Only a small amount of plant tissue is needed for this method (on the order of μg) and an accurate mass spectrometer. Other measures of WUE using gas exchange data such as A/E (i.e., assimilation rate/transpiration) are most useful when measured repeatedly over time (e.g., diurnal) because they are vulnerable to changes in environmental conditions such as light, humidity, and soil conditions. Carbon isotopes have been used successfully to assess drought tolerance in other legume crops such as chickpea, but it is not yet a ubiquitous tool among pigeonpea programs (Kashiwagi et al. 2006).

Whole plant water movement is a function of the water potential gradient from the soil, roots, plant leaves, and atmosphere (often referred to SPAC). Water potential in leaf tissue has three main components: gravity potential, osmotic potential, and turgor potential (the turgor potential component is replaced with a pressure potential term when assessing xylem water potentials). Comparative studies have shown pigeonpea maintains a higher leaf water status during times of terminal drought, allowing for prolonged survival in soils with low water potentials (Choudhary et al. 2011). This may be attributed to pigeonpea's ability to osmotically adjustment tissues to maintain a more favorable water status (e.g., Flower and Ludlow 1986, 1987). However, it is important to note that osmotic adjustment is an active process (requiring energy), in contrast to the passive change in solute concentrations that can result from a change in plant water status. Although plants must invest energy to synthesize organic molecules and shuttle ions across cell membranes, a higher capacity for osmotic adjustment enhances tolerance to osmotic stress and can prolong survival (Mao et al. 2009). This can be achieved through the accumulation of a range of small osmolytes, such as proline, malate, and other small metabolites.

Osmotic adjustment is involved in another major abiotic stress, salinity and is very similar to drought in that poses an osmotic challenge to plants (Munns and Tester 2008). Cultivars and wild *Cajanus* have been observed to vary in their salinity tolerance (e.g., Johansen et al. 1988; Subbarao et al. 1990, 1991; Srivastava et al. 2006). Salinity can limit germination, plant vigor, and yield of agricultural crops especially in arid and semiarid regions (Munns and Tester 2008; Latef 2014; Aggarwal et al. 2012; Porcel et al. 2012). Approximately 20% of irrigated land worldwide currently is affected by salinity, particularly in arid, and desert lands, which comprise 25% of the total land area of our planet (Yeo 1999). High salinity affects plants in several ways: water stress, ion toxicity, nutritional disorders, oxidative stress, alteration of metabolic processes, membrane disorganization, reduction of cell division and expansion, and genotoxicity (Hasegawa et al. 2000; Munns 2002; Djanaguiraman and Prasad 2013). Together these effects significantly reduce plant growth, development, and survival (Hameed et al. 2014).

Several criteria have been used in screening for salinity tolerance including germination, radicle length, dry weight production, shoot length, cell survival, plant biomass, nodulation, number of pods, grain yield, and K+/Na+ ratio (Toker and Mutlu 2011). However, there are physiological tools available that can accurately determine how a plant accumulates osmolytes and the relationship between water content and water stress, i.e., the pressure-volume curve. The pressure-volume curve can be generated with a Scholander pressure chamber and a balance and is plotted with the inverse water potential on the *y*-axis, and the deficit from saturated water content on the *x*-axis from repeatedly taking measures as the plant tissue dries. From the resulting curve, several parameters can be extracted such as the osmotic potential, turgor loss point, and bulk elastic modulus (Tyree and Hammel 1972). The pressure-volume curve technique has been used in several other crop systems (e.g., Katerji et al. 1997, 1998) but is not yet integrated into pigeonpea programs despite past interest (i.e., Flower and Ludlow 1986, 1987). This physiological technique is ideal to assess plant water status, capacity for osmotic adjustment, and thus to identify these favorable traits. These osmotic adjustment traits can then be used in QTL analysis to further breeding efforts (e.g., Teulat et al. 1998). Greater integration of physiological and genetic tools is needed in pigeonpea agroecology and breeding programs.

Classic morphological traits associated with drought tolerance can also be selective targets. Due to pigeonpea's generally deep root system, it is considered a drought tolerant crop because having a deeper root system allows access water at greater depths in the soil. Onim (1983) observed differences in root length that was positively correlated with seed yield among genotypes in pigeonpea; therefore, it seems root characteristics and architecture are posed as ideal candidate traits for selection in breeding programs for water-limited environments. This can be aided by root optimization models which can be used to predict and assess plant responses to environmental factors (Dunbabin et al. 2002; Pages et al. 2004). Our understanding of how root architecture contributes to drought tolerance is rapidly growing (e.g., Lynch et al. 2014; Lynch 2015) and can be applied to developing more drought resilient root systems for crops like pigeonpea. Furthermore, some studies classify the variation in pigeonpea life histories into 4 groups of differing life histories regarding time to maturity: 90–120 days, 120–150 days, 150–200 days, and 200–300 days (Choudharry et al. 2011). If these genotypes are made a part of the germplasm collection, one clear strategy to improve yield under water limiting environments is to match genotypes with seasonal patterns, although some research suggests shorter period genotypes experience greater deficits in seed yield under stress (Lopez et al. 1994). Collection of landraces or wild crop relatives of pigeonpea with locally adapted genotypes in harsh environments can be used to create a comprehensive germplasm collection that affords plant breeders access to desirable traits such as extensive root systems, phenology, water-use efficiency, osmotic adjustment, etc. However, current collections lack sufficient passport data or depth of collection to thoroughly explore local adaptation (Jaggal et al. 2014). Ideally, large collections spanning ecological gradients allow one to find accessions from regions with climates similar to areas that are either now marginal or that are predicted to be marginal with future climate change. Using material from these marginal climates increases the chances it harbors adaptive variation to these conditions. Without large collections, spanning ecological gradients, assessing local adaptation, and utilizing it for breeding in wild relatives are not feasible. Most collections predate GPS technology and lack precise site of collection descriptions. Furthermore, although landrace collections, particularly at ICRISAT, are large (see below), wild relative collections, which harbor higher genetic variation, are limited, as is the case in most other crops (Maxted and Kell 2012).

7.2.2 Biotic Challenges to Pigeonpea Production

Pigeonpea suffers from a range of biotic stresses, from nearly 200 herbivorous insects, nematodes, fungal, bacterial, and viral pathogens. The major diseases affecting pigeonpea production are Fusarium wilt (FW), sterility mosaic virus, Phytophthora blight disease, and major pests causing severe damage are pod borer (*Helicoverpa armigera* and *Maruca vitrata*) and pod fly (*Melanagromyza obtusa*) (Minja et al. 2000). Minor diseases include *Macrophomina* stem canker and yellow mosaic virus (in Asia), leaf spot and powdery mildew (in Africa), and witches broom (in the tropical Americas)

(Kannaiyan et al. 1984). The major diseases and pests can cause substantial harm, particularly for small holder farmers with limited access to pesticides or other inputs. Many of these diseases and pests are badly understudied, particularly outside India. For all these challenges, phenotyping remains a laborious task. Genetics sources of resistance are particularly important to meet these challenges.

Helicoverpa armigera pod boring moths cause massive economic damage in pigeonpea, as well as occasional total crop loss (Minja 2001). The moth can destroy entire seeds, as well as leave partially damaged seeds unsuitable for sale or consumption. Pod sucking Hemipterans of several genera can cause equivalent or greater losses in sub-Saharan and south Asian production zones (Minja 2001). These pests can reduce nutrient accumulation in pods and render seeds unfit for market consumption. Seed feeding dipteran larvae of the species *Melanagromyza chalcosoma* also cause substantial losses, although usually less than the other two types of pod-attacking insects. A variety of management strategies such as altered planting dates or harvest techniques can reduce the severity of these pests. Some pesticides are effective, although the different biology of these classes of pod-attacking insects means that no one weapon is likely to be effective. Some biological agents, such as Helicoverpa nuclear polyhedrosis virus (NPV) can be effective but are hard to develop.

Although management can be effective, genetic resistance is very desirable. Cultivated lines and the wild *Cajanus* differ greatly in resistance, with the wild species forming a particularly useful but under-explored reservoir of resistance to insect pest (Shanower et al. 1997; Green et al. 2002; Chougule et al. 2003; Aruna et al. 2005; Mallikarjuna et al. 2007; Sujana et al. 2008; Sharma et al. 2009). Combinations of amylase and protease inhibitors in pigeonpea seeds can reduce feeding by and induce mortality in *Helicoverpa* (Giri and Kachole 1998), and phenolic compounds and morphological structures like trichomes can contribute to deterrence (Green et al. 2003; Sharma et al. 2009). Larger collections of wild material would likely uncover segregating polymorphisms in resistance to these pests, potentially from distinct loci to those segregating in the cultivated material. Furthermore, the increased examination is needed to determine the extent to which genetic resistance to once pest modulates resistance to the other pod-eating pests.

Wild material is a similarly useful source of resistance to diseases, particularly when coupled with molecular tools facilitated by the pigeonpea genome project to identify resistance genes. Pigeonpea mosaic sterility virus is a vectored disease transmitted by the mite *Aceria cajani* (Jones et al. 2004). Because the disease causes pigeonpea to not flower, it is a green plague. Only found in India, it is perhaps the most damaging of the diseases, although all are problematic. Kulkarni et al. (2003) described broad based resistance to *Fusarium* in *Cajanus scarabaeoides,* a species in the secondary gene pool. Jones et al. (2004) described similar mosaic sterility virus resistance in *C. scarabaeoides.* Gangwar and Bajpai (2008) also found mosaic sterility virus resistance from *C. scarabaeoides* and *C. acutifoloius* to segregate in interspecific crosses with cultivated pigeonpea. Since the publication of the pigeonpea genome, some resistance loci have been identified (Singh et al. 2015).

7.2.3 Economic Traits in Wild Relatives of Pigeonpea

Abiotic (e.g., drought and salinity) and biotic (e.g., diseases and pests) stresses constrain and adversely affect pigeonpea production and cause huge economic damage. Breeding strategies to tackle these problems in pigeonpea have been attempted by various researchers (reviewed by Saxena 2008). The breeding programs for developing disease resistant cultivars using resistance gene sources from cultivated pigeonpea germplasm did not succeed in controlling devastating pests (e.g., pod borer). The cultivated gene pool has low genetic polymorphism and lacks resistance alleles to a number of diseases (e.g., Kassa et al. 2012; Saxena et al. 2014)

Alternative approaches of utilizing wild species as a source of resistance have shown promising results, as there are wider genetic diversity and the presence of resistance genes in the wild gene pool.

Wild species have coexisted with pests and pathogens on an evolutionary time scale, and they have developed alleles conferring pest and pathogen resistance (Acosta-Gallegos et al. 1998). These natural defense mechanisms for diseases and pests have been lost during domestication, and intense selection for agriculturally desirable traits such as high yield improved nutritional quality and other desirable agronomic traits. Most wild species have unique traits (e.g., the presence of trichomes) that confer resistance to these diseases and pests (Aruna et al. 2005). Wild species in the secondary and tertiary gene pools also possess useful genes for extra-early flowering and maturity, photoperiod insensitivity, good flowering and pod setting, true annuality, rapid seedling growth (Mallikarjuna and Moss 1995). Wild relatives of pigeonpea are also genetic sources for salinity tolerance (Subbarao 1988; Srivastava et al. 2006), drought tolerance, resistance to sterility mosaic virus, Phytophthora blight disease (Reddy et al. 1996; Mallikarjuna et al. 2005, 2006), tolerance to pod borers (*Helicoverpa armigera* and *Maruca testulalis*), and pod fly (Melanagromyza obtusa) (Saxena 2008). Species in the secondary gene pool such as *Cajanus albicans*, *C. lineatus*, *C. scarabaeoides*, *C. sericeus* have genes for high seed protein, and *Cajanus sericeus* has genes for resistance to sterility mosaic virus and P2 race of Phytophthora blight disease (Saxena 2008). Wild relatives of pigeonpea such as *Cajanus scarabaeoides*, *C. sericeus*, *C. acutifolius*, *C. albicans*, *Rhynchosia aurea*, *R. bracteata,* and *Flemingia bracteata* have shown high resistance to pod borer. *Cajanus platycarpus* has shown resistance to the most virulent race of phytophthora blight disease and the only source of resistance to the P3 race of Phytophthora blight disease (Saxena 2008). Some of the wild relatives of pigeonpea have shown a high level of resistance to pod fly (*Melanagromyza obtusa*) and pod wasp (*Tanaostigmodes cajaninae*) (Sharma et al. 2003).

7.3 Cytoplasmic Male Sterility System in Pigeonpea Breeding

All wild species of *Cajanus* have the same number of chromosomes (2n = 22) and similar karyotype and their interspecific hybrids showed chromosomal homology and complete chromosomal pairing. Efforts have been made to utilize desirable genes of the wild relatives in pigeonpea breeding and improvement programs (Ariyanayagam et al. 1995; Saxena et al. 1996; Tikka et al. 1997; Wanjari et al. 2000; Saxena and Kumar 2003, 2010; Saxena et al. 2005a). Wild species of pigeonpea have been utilized in breeding programs to develop cleistogamous lines (Saxena et al. 1992a, 2010), genetic dwarfs (Saxena and Sharma 1995), and cytoplasmic male sterile (CMS) lines (Saxena 2006). Breeders at ICRISAT have been utilizing the partial out-crossing nature of pigeonpea to develop a high yielding hybrid cultivar (Reddy et al. 1978). Saxena et al. (1992b) developed a pigeonpea hybrid ICPH 8, which showed increased yield gain (30.5%) over the best performing pure line control. Despite the success of releasing high yielding ICPH8 to farmers, it was not adopted effectively due to the high cost of seed production of the hybrid. This led to the development of the CMS breeding system, which was more efficient in large-scale hybrid seed production. Cytoplasmic male sterility (CMS) is a phenotypic expression of incompatibility between nuclear and cytoplasmic genomes and is a maternally inherited trait that has been successfully used as an efficient pollination control system in developing hybrid seed production (Havey 2004). In most cases, CMS is caused by the interaction between the recessive nuclear genes and specific genetic factors housed in the mitochondrial genome which cause dysfunctionality of the anthers and result in male sterility. Fertility can be restored if dominant genes substitute the recessive nuclear genes or fertility-inducing factors arise in the mitochondrial genome. Three parents are required to maintain the CMS-based breeding system: a male sterile line (A-line), maintainer line (B-line), and fertility restorer line (R-line). CMS systems can be caused by spontaneous mutation, intraspecific, interspecific, or intergeneric crosses. About 75% of the CMS systems are a result of interspecific and intergeneric crosses (Kaul 1988). The absence of CMS lines within pigeonpea germplasm led to the synthesis of CMS lines through interspecific crosses between cultivated pigeonpea and wild species using the cytoplasm genome of the wild parent and nuclear genome of the cultivated parent (Saxena 2008). To date, seven CMS breeding lines (A_1–A_7) have been developed by crossing wild parental lines with the cultivated pigeonpea parent. However, all except one failed to use in a commercial seed production system. The exception is the CMS system (A_4) that uses the wild progenitor species *Cajanus cajanifolius* cytoplasm (Saxena et al. 2005b). High yielding and stable seeds are produced and are being used extensively by breeders to develop commercial pigeonpea hybrids.

7.4 Hybrid Vigor in Pigeonpea

For the past five decades, breeders, particularly in India, have actively pursuing pigeonpea improvement with the objective of developing high yielding varieties. Though over 100 new cultivars were released, crop productivity remains low and stagnant at around 750 kg/ha (Saxena and Sawargaonkar 2014). However, recent progress made on hybrid technology in pigeonpea resulted in high yielding hybrid cultivars. Currently, efforts are underway to identify diverse heterotic groups in pigeonpea with the objective of developing exceptionally high yielding hybrids using a diverse set of elite and hybrid parents (Saxena and Sawargaonkar 2014).

After decades of intense research, the world's first commercial CMS-based legume (pigeonpea) ("magic pea hybrid") was developed by ICRISAT. This breakthrough was achieved through a CMS system of using the wild progenitor of pigeonpea, *Cajanus cajanifolius*, as a parental line (Saxena 2009). Unlike previous attempts of hybrid development (ICPH 8) in pigeonpea, three CMS hybrids were developed and released. These are ICPH 2671 (Saxena et al. 2013a), ICPH 2740 (Saxena and Sawargaonkar 2014), and ICPH 3762 (Pazhamala et al. 2015), which are stable across diverse environments and had an excellent male fertility restorer system. The hybrids give about 40% yield increase under farmers' field over the best controls and showed resistance to wilt and sterility mosaic diseases (Saxena 2009; Saxena and Sawargaonkar 2014). To augment the CMS technology, genomic analysis of the mitochondrial genome of the three lines (CMS, maintainer, and restorer lines) is being pursued (Pazhamala et al. 2015). Additionally, cross-applicable markers for testing pigeonpea hybrid purity are already developed (Bohra et al. 2014).

7.4.1 Molecular Tools and Trait Discovery

Development of new genomic tools and resources has brought a revolution in plant breeding. It enables deciphering and genetic dissection of traits of importance. Next generation sequencing (NGS) technologies coupled with bioinformatics advances are availing abundant genomic information of many crops. Discovering of new genes, QTLs, and molecular markers associated with important agronomic traits is now becoming routine (Varshney et al. 2009, 2015; Varshney 2016).

In pigeonpea, rapid progress has been made recently, and abundant genomic resources are currently available. This will speed up trait discovery and promote molecular plant breeding. Molecular markers starting from the classical gel and hybridized-based technologies (e.g., RAPD, RFLP, AFLP, DArT) to high throughput SNP markers have been developed for pigeonpea over the years (see review, Pazhamala et al. 2015). Moreover, a draft genome sequence of pigeonpea has been developed (Varshney et al. 2012a) and is the basis for future advances. Recently, efforts have been made to identify genes and QTLs associated with important agronomic traits such as drought tolerance (Varshney et al. 2012a), plant height and earliness (Kumawat et al. 2012), and determinacy (Mir et al. 2014).

Recently, increasing the effort has been dedicated to building the types of complex populations that are needed to take advantage of the depth of data provided by next generation sequencing approaches. One approach is to establishing genome-wide association study panels for gene discovery across diverse germplasm. Some early success with GWAS has already occurred (e.g., Huang and Han 2013; Korte and Farlow 2013). In addition to GWAS, complex crossing designs can combine the benefits of GWAS with QTL. A nested association mapping (NAM) population for pigeonpea has been built (Yu et al. 2008; McMullen et al. 2009). These populations are an excellent tool for uncovering the genetic basis of traits because they use widely divergent parents and bring to bear the combined benefits of both QTL mapping and genome-wide association. A related tool, a multi-parent advanced generation intercross (MAGIC) population, has also been developed for pigeonpea (Cavanagh et al. 2008). Although MAGIC populations generally have fewer parents than NAM populations and reduced capacity for gene discovery, if they are set up with elite parents they will more quickly yield progeny suitable for breeding programs. Both sets of tools are ultimately useful and important. Full genome sequencing of parents of many of the lines used in these populations will further facilitate their use (Kumar et al. 2016). In addition to these critical population tools, we encourage the more widespread use of advance backcross introgression lines to effectively capture variation in wild germplasm (e.g., Tanksley and McCouch 1997; Warschefsky et al. 2014). Although we are not aware of such efforts, we see great potential in these crossing designs for the effective use of wild germplasm.

7.4.2 Conclusion

Germplasm resources remain a critical component of all breeding programs. At a time of rapid habitat loss in all ecosystems, and a time of declining funding for many gene banks, and germplasm centers, germplasm conservation is of increasing importance. The risk of loss of many collections, such as that at the Vavilov Institute for Research in St. Petersberg, Russia, remains high. Many other collections, such as all those in the Consultative Group on International Agricultural Research and the USDA have unstable funding sources. And wild populations of all species, such as *Cajanus* species in South Asia, have declining habitat areas and new threats from invasive species and land conversion. In the face of these risks, steps are required to better protect germplasm.

Acknowledgements The authors thank the editors and one anonymous reviewer for helpful comments on an earlier draft of the manuscript. EvW is supported by the Feed the Future Chickpea Innovation Laboratory for Climate Resilient Chickpea (AID-OAA-A-14-00008), by NSF PGRP 1339346, by a grant from the Global Crop Diversity Trust CWR14NOR2 3.3, by HHMI award no. 52006924 to Florida International University's discipline-based education research group, and USDA-NIFA Hispanic Serving Institutions Grant 2011-38422-30804 to FIU's agroecology program. The Global Crop Diversity Trust activity is a part of the initiative "Adapting Agriculture to Climate Change: Collecting, Protecting and Preparing Crop Wild Relatives" which is supported by the Government of Norway. The project is managed by the Global Crop Diversity Trust with the Millennium Seed Bank of the Royal Botanic Gardens, Kew and implemented in partnership with national and international gene banks and plant breeding institutes around the world. For further information, see the project website: http://www.cwrdiversity.org/.

References

Acosta-Gallegos JA, Quintero C, Vargas J, Toro O, Tohme J, Cardona C (1998) A new variant of arcelin in wild common bean, *Phaseolus vulgaris* L. from southern Mexico. Genet Resour Crop Evol 45:235–242

Aggarwal A, Kadian N, Neetu K, Tanwar A, Gupta KK (2012) Arbuscular mycorrhizal symbiosis and alleviation of salinity stress. J Appl Nat Sci 4(1):144–155

Ariyanayagam RP, Rao AN, Zaveri PP (1995) Cytoplasmic-genic male- sterility in interspecific matings of *Cajanus*. Crop Sci 35:981–985

Aruna R, Manohar RD, Reddy LJ, Upadhyaya HD, Sharma HC (2005) Inheritance of trichomes and resistance to pod borer (Helicoverpa armigera) and their association in interspecific crosses between cultivated pigeonpea (*Cajanus cajan*) and its wild relative *C. scarabaeoides*. Euphytica 145:247–257

Bohra A, Singh IP, Yadav AK, Pathak A, Soren KR, Chaturvedi SK et al (2014) Utility of informative SSR markers in the molecular characterization of cytoplasmic genetic male sterility-based hybrid and its parents in pigeonpea. Natl Acad Sci Lett 38:13–19

Brown TA, Jones MK, Powell W, Allaby RG (2009) The complex origins of domesticated crops in the Fertile Crescent. Trends Ecol Evol 24:103–109

Cavanagh C, Morell M, Mackay I, Powell W (2008) From mutations to MAGIC: resources for gene discovery, validation and delivery in crop plants. Curr Opin Plant Biol 11(2):215–221

Chaves M, João PM, João SP (2003) Understanding plant responses to drought—from genes to the whole plant. Funct Plant Biol 30(3):239–264

Choudhary AK, Sultana R, Pratap A, Nadarajan N, Jha UC (2011) Breeding for abiotic stresses in pigeonpea. J Food Legumes 24(3):165–174

Chougule NP, Hivrale VK, Chhabda PJ, Giri AP, Kachole MS (2003) Differential inhibition of Helicoverpa armigera gut proteinases by proteinase inhibitors of pigeonpea (*Cajanus cajan*) and its wild relatives. Phytochemistry 64:681–687

Djanaguiraman M, Prasad PV (2013) Effects of salinity on ion transport, water relations and oxidative damage. In: Ecophysiology and responses of plants under salt stress, Springer, New York, pp 89–114

Donovan LA, Ehleringer JR (1994) Potential for selection on plants for water-use efficiency as estimated by carbon isotope discrimination. Am J Bot 81:927–935

Dunbabin VM et al (2002) Modelling the interactions between water and nutrient uptake and root growth. Plant Soil 239:19–38

Ehlers W, Goss MJ (2003) Water dynamics in plant production. CABI, Wallingford, Oxfordshire, UK

FAO (2012) Online agriculture statistics. http://www.faosat.org

Farooq M et al (2009) Plant drought stress: effects, mechanisms and management. Sustainable Agriculture. Springer, Netherlands, pp 153–188

Farquhar GD, Hubick KT, Condon AG, Richards RA (1989) Carbon isotope fractionation and plant water-use efficiency. Stable isotopes in ecological research. Springer, New York, pp 21–40

Flower DJ, Ludlow MM (1986) Contribution of osmotic adjustment to the dehydration tolerance of water-stressed pigeon pea (Cajanus cajan (L.) millsp.) leaves. Plant, Cell Environ 9:33–40

Flower DJ, Ludlow MM (1987) Variation among accessions of pigeonpea (*Cajanus cajan*) in osmotic adjustment and dehydration tolerance of leaves. Field Crops Res 17:229–243

Ford-Lloyd BV, Schmidt M, Armstrong SJ, Barazani O, Engels J, Hadas R, Hammer K, et al. (2011) Crop wild relatives—undervalued, underutilized and under threat? BioScience 61(7):559–565

Fortunato RH (2000) Systematic relationships in *Rhynchosia* (Cajaninae Phaseoleae–Papilionoideae–Fabaceae) from the neotropics. In: Bruneau A (ed) Herendeen PS. Kew, UK, The Royal Botanic Gardens, pp 339–354

Gangwar LK, Bajpai GC (2008) Screening of interspecific hybrids against sterility mosaic disease in pigeonpea. Legume Res-An Int J 31:306–307

Giri AP, Kachole MS (1998) Amylase inhibitors of pigeonpea (*Cajanus cajan*) seeds. Phytochemistry 47:197–202

Gowda CLL, Upadhyaya HD, Sharma S, Varshney RK, Dwivedi SL (2013) Exploiting genomic resources for efficient conservation and use of chickpea, groundnut, and pigeonpea collections for crop improvement. Plant Genome 6:1–11. doi:10.3835/plantgenome2013.05.0016

Grant OM (2012) Understanding and exploiting the impact of drought stress on plant physiology. In: Abiotic stress responses in plants. Springer, New York, pp 89–104

Green PWC, Stevenson PC, Simmonds MSJ, Sharma HC (2002) Can larvae of the pod-borer, Helicoverpa armigera (Lepidoptera: Noctuidae), select between wild and cultivated pigeonpea Cajanus sp. (Fabaceae)? Bull Entomol Res 92:45–51

Green PW, Stevenson PC, Simmonds MS, Sharma HC (2003) Phenolic compounds on the pod-surface of pigeonpea, *Cajanus cajan*, mediate feeding behavior of Helicoverpa armigera larvae. J Chem Ecol 29:811–821

Hajjar R, Hodgkin T (2007) The use of wild relatives in crop improvement: a survey of developments over the last 20 years. Euphytica 156(1–2):1–13

Harlan JR, de Wet JMJ (1971) Toward a rational classification of cultivated plants. Taxon 20:509–517

Hasanuzzaman M, Nahar K, Alam MM, Roychowdhury R, Fujita M (2013) Physiological, biochemical, and molecular mechanisms of heat stress tolerance in plants. Int J Mol Sci 14(5):9643–9684

Hasegawa PM, Bressan RA, Zhu JK, Bohnert HJ (2000) Plant cellular and molecular responses to high salinity. Ann Rev Plant Biol 51(1):463–499

Havey MJ (2004) The use of cytoplasmic male sterility for hybrid seed production. In: Daniell H, Chase C (eds) Molecular biology and biotechnology of plant organelles. Springer, Dordrecht, The Netherlands, pp 617–628

Holmes MG, Keiller DR (2002) Effects of pubescence and waxes on the reflectance of leaves in the ultraviolet and photosynthetic wavebands: a comparison of a range of species. Plant, Cell & Environ 25 (1):85–93

Huang X, Han B (2014) Natural variations and genome-wide association studies in crop plants. Ann Rev Plant Biol 65:531–551

Jaggal LG, Patil BR, Salimath PM (2014) Evaluation of mini core accessions of pigeonpea (Cajanus cajan (L.) Millsp.) against sterility mosaic disease and fusarium wilt. Karnataka J Agric Sci 27:337–339

Johansen C, Saxena NP, Chauhan YS, Rao GS, Pundir RPS, Rao JVDKK, Jana MK (1988) Genotypic variation in salinity response of chickpea and pigeonpea. In: Proceedings of the International Congress of Plant Physiology, New Delhi, India, 15–20 February 1988

Jones AT, Kumar PL, Saxena KB, Kulkarni NK, Muniyappa V, Waliyar F (2004) Sterility mosaic disease—the "green plague" of pigeonpea: advances in understanding the etiology, transmission and control of a major virus disease. Plant Disease 88(5):436–445

Kannaiyan J, Nene YL, Reddy MV, Ryan JG, Raju TN (1984) Prevalence of pigeonpea diseases and associated crop losses in Asia, Africa and the Americas. Int J Pest Manage 30:62–72

Kashiwagi J, Krishnamurthy L, Singh S, Gaur PM, Upadhyaya HD, Panwar JDS, Basu PS, Ito O, Tobita S (2006) Relationship between transpiration efficiency and carbon isotope discrimination in chickpea (*Cicer arietinum* L.). J SAT Agric Res 2:1–3

Kassa MT, Varma Penmetsa R, Carrasquilla-Garcia N, Sarma BK, Datta S et al (2012) Genetic patterns of domestication in pigeonpea (Cajanus cajan (L.) Millsp.) and wild Cajanus relatives. PLoS ONE 7: e39563

Kassa M, Van der Maesen JG, Krieg C, von Wettberg EJB (2016) Historical and phylogenetic perspectives of pigeonpea. Legume Perspect 11:7–9

Katerji N, Van Hoorn JW, Hamdy A, Mastrorilli M, Karzel EM (1997) Osmotic adjustment of sugar beets in response to soil salinity and its influence on stomatal conductance, growth and yield. Agric Water Manage 34:57–69

Katerji N, Van Hoorn JW, Hamdy A, Mastrorilli M (1998) Response of tomatoes, a crop of indeterminate growth, to soil salinity. Agric Water Manage 38:59–68

Kaul M (1988) Male sterility in higher plants. In: Frankel R, Grossman M, Maliga P (eds) Monographs on theoretical and applied genetics 10. Springer-Verlag, Berlin, pp 775–795

Khoury CK, Castañeda-Alvarez NP, Achicanoy HA, Sosa CC, Bernau V, Kassa MT, Norton SL, van der Maesen LJG, Upadhyaya HD, Ramirez-Vallegas R, Jarvis A, Struik PC (2015) Crop wild relatives of pigeonpea [*Cajanus cajan* (L.) Millsp.]: Distributions, ex situ conservation status, and potential genetic resources for abiotic stress tolerance. BiolConserv 184:259–270

Korte A, Farlow A (2013) The advantages and limitations of trait analysis with GWAS: a review. Plant Methods 9(1):29

Kulkarni NK, Reddy AS, Kumar PL, Vijaynarasimha J, Rangaswamy KT, Muniyappa V, Reddy DVR et al (2003) Broad-based resistance to pigeonpea sterility mosaic disease in accessions of Cajanus scarabaeoides (L.) Benth. Indian J Plant Prot 31:6–11

Kumar V, Khan AW, Saxena RK, Garg V, Varshney RK (2016) First-generation HapMap in *Cajanus* spp. reveals untapped variations in parental lines of mapping populations. Plant Biotechnol J

Kumawat G, Raje RS, Bhutani S, Pal JK, Mithra SVCR, Gaikwad K et al (2012) Molecular mapping of QTLs for plant type and earliness traits in pigeonpea (*Cajanus cajan* L. Millsp.). BMC Genet 13:84

Latef AAHA, Miransari M (2014) The role of arbuscular mycorrhizal fungi in alleviation of salt stress. In: Use of Microbes for the Alleviation of Soil Stresses, Springer, New York, pp 23–38

Lopez FB, Johansen C, Chauhan YS (1994) Limitations to seed yield in short-duration pigeonpea under water stress. Field Crops Res 36:95–102. doi:10.1016/0378-4290(94)90058-2

Lynch JP (2015) Root phenes that reduce the metabolic costs of soil exploration: opportunities for 21st century agriculture. Plant, Cell Environ 38:1775–1784

Lynch JP, Chimungu JG, Brown KM (2014) Root anatomical phenes associated with water acquisition from drying soil: targets for crop improvement. J Exp Bot, eru162

Mallikarjuna N, Moss JP (1995) Production of hybrids between *Cajanus platycarpus* and *Cajanus cajan*. Euphytica 83:43–46

Mallikarjuna N, Deepak J, Reddy MV, Usharani DT (2005) Introgression of Phytophthora blight disease resistance from *Cajanus platycarpus* into short duration pigeonpea [*Cajanus cajan* (L.) Millsp.]. Indian J Genet Plant Breed 65:261–263

Mallikarjuna N, Jadhav D, Reddy P (2006) Introgression of Cajanus platycarpus genome into cultivated pigeonpea. C. cajan. Euphytica 149:161–167

Mallikarjuna N, Sharma HC, Upadhyaya HD (2007) Exploitation of wild relatives of pigeonpea and chickpea for resistance to *Helicoverpa armigera*. J SAT Agric Res 3:4

Mallikarjuna N, Saxena KB, Jadhav DR (2011) Cajanus. In: Kole C (ed) Wild crop relatives: genomic and breeding resources, legume crops and forages. Springer-Verlag, Berlin, Heidelberg, pp 21–33

Mao X, Zhang H, Tian S, Chang X, Jing R (2009) TaSnRK2.4, an SNF1-type serine-threonine protein kinase of wheat (Triticum aestivum L.) confers enhanced multi-stress tolerance in Arabidopsis. J Exp Bot 61:683–696

Maxted N, Kell SP (2009) Establishment of a global network for the in situ conservation of crop wild relatives: status and needs. FAO Commission on Genetic Resources for Food and Agriculture, Rome, Italy, p 266

McMullen MD, Kresovich S, Villeda HS, Bradbury P, Li H, Sun Q, Flint-Garcia S, Thornsberry J, Acharya C, Bottoms C, Brown P (2009) Genetic properties of the maize nested association mapping population. Science 325(5941):737–740

Minja EM (2001) Yield losses due to field pests and integrated pest management strategies for pigeonpe—A synthesis. In Status and potential of pigeonpea in Eastern and Southern Africa: proceedings of a regional workshop, 12–15 Sep 2000, Nairobi, Kenya. B-5030 Gembloux, Belgium: Gembloux Agricultural University; and Patancheru 502 324, Andhra Pradesh, India: International Crops Research Institute for the Semi-Arid Tropics. 232 pp (p. 48). ISBN 92-9066-432-0

Minja EM, Shanower TG, Silim SN, Karuru O (2000) Efficacy of different insecticides for pigeonpea pest management in Kenya. ICPN 7:30–43

Mir RR, Kudapa H, Srikanth S, Saxena RK, Sharma A, Azam S et al (2014) Candidate gene analysis for determinacy in pigeonpea (*Cajanus* spp.). Theor Appl Genet 127:2663–2678

Munns R (2002) Salinity, growth and phytohormones. Salinity: environment—plants—molecules. Kluwer Academic Publishers, Dordrecht, pp 271–290

Munns R, Tester M (2008) Mechanisms of salinity tolerance. Annu Rev Plant Biol 59:651–681

Olsen KM, Wendel JF (2013) A bountiful harvest: genomic insights into crop domestication phenotypes. Annu Rev Plant Biol 6447–70

Onim JFM (1983) Association between grain yield and drought resistance in marginal rainfall areas in Kenya. In: Holmes JC, Tahor WM (eds) More food from better technology, pp 864–872

Pages L et al (2004) Root Typ: a generic model to depict and analyse the root system architecture. Plant Soil 258:103–119

Pazhamala L, Saxena RK, Singh VK, Sameerkumar CV, Kumar V, Sinha P et al (2015) Genomics-assisted breeding for boosting crop improvement in pigeonpea (*Cajanus cajan* L.). Front. Plant Sci 6:50

Porcel R, Aroca R, Ruiz-Lozano JM (2012) Salinity stress alleviation using arbuscular mycorrhizal fungi. A review. Agron Sustain Dev 32(1):181–200

Purugganan MD, Fuller DQ (2009) The nature of selection during plant domestication. Nature 457:843–848

Reddy BVS, Green JM, Bisen SS (1978) Genetic male-sterility in pigeonpea. Crop Sci 18:362–364

Reddy MV, Raju TN, Sheila VK (1996) Phytophthora blight disease in wild pigeonpea. ICPN 3:52–53

Saxena KB (2006) Hybrid pigeonpea seed production manual. International Crops research Institute for the Semi Arid Tropics. Patancheru, India. Bulletin 74: 27

Saxena KB (2008) Genetic improvement of pigeonpea—a review. Trop Plant Biol 1:159–178

Saxena KB (2009) Evolution of hybrid breeding technology in pigeonpea. In: Ali M, Kumar S, (eds.) Milestones in food legume research. Indian Institute of Pulses Research, Kanpur, India. 82–114

Saxena KB, Kumar RV (2003) Development of a cytoplasmic- nuclear male-sterility system in pigeonpea using C. scarabaeoides (L.) Thouars. Indian J Genet 63:225–229

Saxena KB, Kumar RV (2010) Insect-aided natural out-crossing in four wild relatives of pigeonpea. Euphytica 173:329–335

Saxena KB, Sawargaonkar SL (2014) First information on heterotic groups in pigeonpea [Cajanus cajan (L.) Millsp.]. Euphytica 200:187–196

Saxena KB, Sharma D (1995) Sources of dwarfism in pigeonpea. Indian J Pulses Res 8:1–6

Saxena KB, Singh L, Gupta MD (1990) Variation for natural outcrossing in pigeonpea. Euphytica 46:143–148

Saxena KB, Singh L, Ariyanayagam RP (1992a) Role of partial cleistogamy in maintaining genetic purity of pigeonpea. Euphytica 66:225–229

Saxena KB, Chauhan YS, Johansen C, Singh L (1992b) Recent developments in hybrid pigeonpea research. New Frontiers in Pulses Research and Development. Indian Institute of Pulses Research, Kanpur, India, pp 58–69

Saxena KB, Rao AN, Singh U, Remnanadan P (1996) Interspecies variation in *Cajanus* platicarpus for some agronomic traits and crossability. ICPN 3:49–51

Saxena KB, Srivastava DP, Chauhan YS, Ali M (2005a) Hybrid pigeonpea. In: Ali M, Kumar S (eds) Advances in pigeonpea research. IIPR Kanpur, India, pp 96–133

Saxena KB, Kumar RV, Srivastava N, Bao S (2005b) A cytoplasmic-nuclear male-sterility system derived from a cross between Cajanus cajanifolius and Cajanus cajan. Euphytica 145:289–294

Saxena KB, Sultana R, Mallikarjuna N, Saxena RK, Kumar RV, Sawargaonkar SL, Varshney RK (2010) Male-sterility systems in pigeonpea and their role in enhancing yield. Plant Breed 129:125–134

Saxena RK, Von Wettberg E, Upadhyaya HD, Sanchez V, Songok S, Saxena K, Varshney RK (2014) Genetic diversity and demographic history of Cajanus spp. illustrated from genome-wide SNPs. PLoS ONE 9:e88568

Schuster WS, Sandquist DR, Phillips SL, Ehleringer JR (1992) Comparisons of carbon isotope discrimination in populations of aridland plant species differing in lifespan. Oecologia 91:332–337

Shanower TG, Yoshida M, Peter JA (1997) Survival, growth, fecundity, and behavior of Helicoverpa armigera (Lepidoptera: Noctuidae) on pigeonpea and two wild Cajanus species. J Econ Entomol 90:837–841

Sharma HC, Pampapathy G, Reddy LJ (2003) Wild relatives of pigeonpea as a source of resistance to the pod fly (Melanagromyza obtusa Malloch) and pod wasp (Tanaostigmodes cajaninae La Salle). Genet Resour Crop Evol 50:817–824

Sharma HC, Sujana G, Rao DM (2009) Morphological and chemical components of resistance to pod borer, *Helicoverpa armigera* in wild relatives of pigeonpea. Arthropod-Plant Interact 3:151–161

Singh N, Tyagi RK, Pandey C (2013) Genetic resources of pigeonpea: conservation for use. National Bureau of Plant Genetic Resources (NBPGR), New Delhi, pp 1–49

Singh M, Gautam NK, Rana MK, Dahiya OP, Dutta M, Bansal KC (2014) Pigeonpea genetic resources and its utilization in India, current status and future prospects. J Plant Sci Res 1:107

Singh VK, Khan AW, Saxena RK, Kumar V, Kale SM, Sinha P, et al Kumar, S (2015) Next-generation sequencing for identification of candidate genes for Fusarium wilt and sterility mosaic disease in pigeonpea (*Cajanus cajan*). Plant Biotechnol J

Srivastava N, Vadez V, Upadhyaya HD, Saxena KB (2006) Screening for intra and inter specific variability for salinity tolerance in pigeonpea (Cajanus cajan (L.) Millsp.) and its related wild species. E-J SAT Agric Res Crop Improv 2:1

Subbarao GV (1988) Salinity tolerance in pigeonpea (Cajanus cajan (L.) Millsp.) and its wild relatives. PhD thesis, Indian Inst Technol, Kharagpur, India

Subbarao GV, Johansen C, Jana MK, Rao JK (1990) Physiological basis of differences in salinity tolerance of pigeonpea and its related wild species. J Plant Physiol 137:64–71

Subbarao GV, Johansen C, Jana MK, Rao JVDK (1991) Comparative salinity responses among pigeonpea genotypes and their wild relatives. Crop Sci 31:415–418

Sujana G, Sharma HC, Rao DM (2008) Antixenosis and antibiosis components of resistance to pod borer Helicoverpa armigera in wild relatives of pigeonpea. Int J Trop Insect Sci 28:191–200

Tang H, Sezen U, Paterson AH (2010) Domestication and plant genomes. Curr Opin Plant Biol 13:160–166

Tanksley SD, McCouch SR (1997) Seed banks and molecular maps: unlocking genetic potential from the wild. Science 277(5329):1063–1066

Teulat B, This D, Khairallah M, Borries C, Ragot C, Sourdille P, Charrier A et al (1998) Several QTLs involved in osmotic-adjustment trait variation in barley (*Hordeum vulgare* L.). Theor Appl Genet 96:688–698

Tikka SBS, Parmar LD, Chauhan RM (1997) First record of cytoplasmic-genic male sterility in pigeonpea (Cajanus cajan (L.) Millsp.) through wide hybridization. GAU Res J. 22:160–162

Toker C, Mutlu N (2011) 16 Breeding for abiotic stresses. In: Pratap A, Kumar J (eds) Biology and breeding of food legumes, CABI Walingford, Oxfordshire, UK, p 241

Toker C, Lluch C, Tejera NA, Serraj R, Siddique KHM (2007) 23 Abiotic Stresses. Chickpea breeding and management. Chickpea breeding and management, CABI Walingford, Oxfordshire, UK, p 474

Tyree MT, Hammel HT (1972) The measurement of the turgor pressure and the water relations of plants by the pressure-bomb technique. J Exp Bot 23(1):267–282

Upadhyaya HD, Reddy KN, Gowda CLL, Singh S (2007) Phenotypic diversity in the pigeonpea (Cajanus cajan) core collection. Genet Resour Crop Evol 54:1167–1184

Upadhyaya HD, Reddy KN, Shivali S, Varshney RK, Bhattacharjee R, Singh S et al (2011) Pigeonpea composite collection for enhanced utilization of germplasm in crop improvement programs. Plant Genet Resour 9:97–108

Upadhyaya HD, Reddy KN, Singh S, Gowda CLL (2013) Phenotypic diversity in Cajanus species and identification of promising sources for agronomic traits and seed protein content. Genet Resour Crop Evol 60:639–659

Van der Maesen LJG (1986) Cajanus DC and Atylosia W and A (Leguminosae). Agricultural University Wageningen papers 85–4:1–225

Varshney RK (2011) Application of next generation sequencing and genotyping technologies to develop large-scale genomic resources in SAT legume crops. In: Acharya NG (ed) National symposium on genomics and crop improvement: relevance and reservations, Ranga Agricultural University, Rajendranagar, Hyderabad, 25–27 Febrauary 2010

Varshney RK (2016) Exciting journey of 10 years from genomes to fields and markets: some success stories of genomics-assisted breeding in chickpea, pigeonpea and groundnut. Plant Sci 242:98–107

Varshney RK, Nayak SN, May GD, Jackson SA (2009) Next-generation sequencing technologies and their implications for crop genetics and breeding. Trends Biotechnol 27:522–530

Varshney RK, Penmetsa RV, Dutta S, Kulwal PL, Saxena R, Datta S et al (2010) Pigeonpea genomics initiative (PGI): an international effort to improve crop productivity of pigeonpea (*Cajanus cajan* L.). Mol Breed 26:393–408

Varshney RK, Chen W, Li Y, Bharti AK, Saxena RK, Schlueter JA et al (2012) Draft genome sequence of pigeonpea (*Cajanus cajan*), an orphan legume crop of resource-poor farmers. Nat Biotechnol 30:83–89

Varshney RK, Kudapa H, Pazhamala L, Chitikineni A, Thudi M, Bohra A, Ellis N et al (2015) Translational genomics in agriculture: some examples in grain legumes. Crit Rev Plant Sci 34:169–194

Vaughan DA, Balazs E, Heslop-Harrison JS (2007) From crop domestication to super-domestication. Ann Bot 100:893–901

Vavilov NI (1922) The law of homologous series in variation. J Genet 12(1):47–89

Wanjari KB, Patil AN, Patel MC, Manjaya JG (2000) Male sterility derived from *Cajanus sericeus* and *Cajanus cajan*: Confusion of cytoplasmic male sterility with dominant genic male-sterility. Euphytica 115:59–64

Warschefsky E, Penmetsa RV, Cook DR, von Wettberg EJ (2014) Back to the wilds: tapping evolutionary adaptations for resilient crops through systematic hybridization with crop wild relatives. Am J Bot 101 (10):1791–1800

Yeo AR, Flowers SA, Rao G, Welfare K, Senanayake N, Flowers TJ (1999) Silicon reduces sodium uptake in rice (Oryza sativa L.) in saline conditions and this is accounted for by a reduction in the transpirational bypass flow. Plant, Cell & Environ 22(5):559–565

Yoo CY et al (2009) Regulation of transpiration to improve crop water use. Crit Rev Plant Sci 28:410–431

Yu J, Holland JB, McMullen MD, Buckler ES (2008) Genetic design and statistical power of nested association mapping in maize. Genetics 178(1):539–551

8 Whole-Genome Sequencing of Pigeonpea: Requirement, Background History, Current Status and Future Prospects for Crop Improvement

Rachit K. Saxena and Rajeev K. Varshney

Abstract

Despite of being a very important crop, pigeonpea did not have genomic resources until 2005. Pigeonpea Genomics Initiative (PGI) supported by Indian Council of Agricultural Research (ICAR) under Indo-US Agriculture Knowledge Initiative was the first major initiative that delivered first set of molecular markers in large numbers, first set of mapping populations, first set of transcriptome assemblies, etc. Subsequently, two consortia—1) International Initiative for Pigeonpea Genomics (IIPG), led by International Crops Research Institute for the Semi-Arid Tropics (ICRISAT) and 2) Led by National Research Centre on Plant Biotechnology (NRCPB)—delivered two draft genome assemblies for Asha (ICPL 87119) variety. In summary, all these genomic resources transformed pigeonpea from an 'orphan crop' to 'genomics resources-rich crop'. After publication of draft genome sequences, a detailed plan was developed to utilize draft genome information for pigeonpea improvement. This plan in the form of a proposal was approved by Ministry of Agriculture, Government of India and United States Agency for International Development (USAID)—India. In addition to this major project, two additional projects were funded by Department of Biotechnology, Government of India. All these efforts have established high-density genotyping platforms such as genotyping by sequencing (GBS) and 'Axiom® *CajanusSNP* Array', produced the first generation HapMap, generated whole-genome re-sequencing data of >400 pigeonpea lines, evaluated several mapping populations for desired traits, established marker–trait association for several traits of interest to breeders and also identified best-performing lines. Additionally, multi-parent advance generation inter-cross (MAGIC) and nested association mapping

R.K. Saxena (✉) · R.K. Varshney
International Crops Research Institute for the Semi-Arid Tropics (ICRISAT), Patancheru 502324, India
e-mail: r.saxena@cgiar.org

R.K. Varshney et al. (eds.), *The Pigeonpea Genome*, Compendium of Plant Genomes, DOI 10.1007/978-3-319-63797-6_8

(NAM) populations are being developed. With the availability of above-mentioned information, next few years will be witnessing application of genomics-assisted breeding for pigeonpea improvement. It is anticipated that improved pigeonpea lines developed through genomics interventions will reach to farmers' fields and elevate the game towards pulse sufficiency for poor farmers in arid and semi-arid regions of the world in near future.

8.1 Introduction

Much has been written on the importance of pigeonpea in previous chapters of this book and in many other published articles. However, until recently very few breeding programs have remained engaged in systematic pigeonpea research involving modern biology approaches. Moreover, the genetic material used in these breeding programs have limited genetic base. There is a need to diversify the genetic base and deploy modern approaches such as genomics-assisted breeding (GAB) for pigeonpea improvement. GAB has helped development of superior varieties and hybrids in several crops, in cereals (Varshney et al. 2005, 2006, 2010a) and legumes such as chickpea (Varshney et al. 2013) and groundnut (Varshney et al. 2014). The pigeonpea crop, however, remained untouched by the genomics revolution until 2005 and resulted as an 'orphan crop legume'. To enable GAB in this crop, the Pigeonpea Genomics Initiative (PGI) was started in late 2006 with financial support from Indian Council of Agricultural Research (ICAR) under the umbrella of Indo-US Agricultural Knowledge Initiative (AKI), US National Science Foundation's Plant Genome Research Program and the Generation Challenge Program (GCP). As a result of intensive efforts in PGI and several other programmes, a significant amount of genomic resources such as molecular markers, mapping populations and genetic maps was developed in pigeonpea (Varshney et al. 2010b). These collaborative efforts significantly benefited pigeonpea research community and transformed an 'orphan crop legume' to 'genomic resources-rich crop'. Although thousands of molecular markers and a number of mapping populations were developed in pigeonpea under PGI, very few genetic maps with low level of marker density and only three quantitative trait loci (QTL)-based studies were conducted (Bohra et al. 2012; Gnanesh et al. 2011; Kumawat et al. 2012). This has happened mainly because of low level of genetic diversity present in the pigeonpea-cultivated gene pool. To overcome the low level of genetic diversity bottleneck in the cultivated gene pool and for deploying GAB, following two options were considered: (1) develop novel genetic material with enhanced genetic diversity and (2) systematically scan entire pigeonpea genome for all possible variations. In PGI, initially a clone-by-clone approach was proposed to sequence the pigeonpea genome. However, due to lack of funds through PGI and availability of low-cost and high-throughput next generation sequencing (NGS) technologies encouraged pigeonpea genomics community for developing genome sequence based on whole-genome shotgun approach. At the later stage, International Initiative for Pigeonpea Genomics (IIPG) (http://ceg.icrisat.org/gt-bt/iipg/Home.html) was floated to sequence genome and developed other genomic resources. As a result of efforts from IIPG and National Research Centre on Plant Biotechnology (NRCPB), two drafts of pigeonpea genome were published in Nature Biotechnology (Varshney et al. 2012) and Journal of Plant Biochemistry and Biotechnology (Singh et al. 2012). Nevertheless, the genomic resources developed in PGI were pivotal in both pigeonpea genome sequencing projects.

8.2 Genome Sequence Initiatives

The year 2012 can be consider a significant year not only for pigeonpea but for the legume community. As mentioned above, IIPG and NRCPB delivered draft pigeonpea genome sequence almost at the same time, one was based on Illumina sequence data (Varshney et al. 2012) and the other one was based on 454 GS-FLX sequence data (Singh et al. 2012). These efforts delivered the genome sequence information of the first grain legume as well as the second food legume, after soybean to help increase the efficiency of pigeonpea improvement by integrating biotechnological tools in conventional breeding and to utilize the genome information of pigeonpea for analysing other legume species. It was also anticipated that the pigeonpea genome information will be useful to understand the genetic basis of hybrid vigour in pigeonpea, develop new methods for hybrid breeding.

Background history and glimpse of the two pigeonpea genome initiatives have been presented below:

8.2.1 Draft Genome Sequence from IIPG (Varshney et al. 2012)

IIPG delivered the genome sequence through partnership of 30 scientists from 12 institutes in 6 countries and published in Nature Biotechnology. Initial partners in the initiative were International Crops Research Institute for the Semi-Arid Tropics (ICRISAT), Patancheru, India; CGIAR Generation Challenge Programme (GCP), Mexico; Beijing Genomics Institute (BGI)-Shenzhen, Shenzhen, China; University of Georgia, Athens, Georgia, USA; National Center for Genome Resources (NCGR), Santa Fe, New Mexico, USA; University of North Carolina, Charlotte, North Carolina, USA; National University of Ireland Galway (NUIG), Botany and Plant Science, Galway, Ireland; University of California, Davis, California, USA; Monsanto Company, Creve Coeur, Missouri, USA; Cold Spring Harbor Laboratory, Cold Spring Harbor, New York, USA; Department of Biology, University of Copenhagen, Denmark; BGI-Americas, Cambridge, Massachusetts, USA. This work received partial funding and in-kind support from CGIAR, GCP, US National Science Foundation, BGI-Shenzhen, China and ICRISAT, India. The decoding of pigeonpea genome was also backed by Indian Council of Agricultural Research (ICAR), India and supported through financial support to some earlier work that was used for analysing genome sequence data.

It is important to mention that above-mentioned partnership was accomplished without any specific funded project for the purpose. The two most important features of this partnership includes following: (1) total expenditure of <US$200 K, whereas in general genome sequencing projects involve millions of US$s, (2) completion of pigeonpea genome sequencing in just about 2 years, unlike many other previous genome sequencing projects (of the same or even smaller genome sizes) generally took 5–10 years or even more (http://www.icrisat.org/newsroom/news-releases/icrisat-pr-2011-media20.htm).

The consortium used ‘ICPL 87119’, popularly known as ‘Asha’ a pigeonpea variety for the generation and analysis of genome sequence. Illumina sequencing technology was used to generate 237.2 Gb of sequence data from 22 different insert size libraries ranging from 180 bp to 20 Kb. This sequence data along with Sanger-based BAC-end sequences and a genetic map assembled ~73% (605.78 Mb) of the 833-Mbp pigeonpea genome. Genome analysis predicted 48,680 genes for pigeonpea and also showed the potential role of some gene families during evolution/domestication, e.g. drought tolerance-related genes. Although a few segmental duplication events were found, recent genome-wide duplication events, such as seen in soybean, were not observed.

8.2.2 Draft Genome Sequence from NRCPB (Singh et al. 2012)

This draft genome, coordinated by NRCPB, was reported in Journal of Plant Biochemistry and Biotechnology. A number of research institutes, namely NRCPB, Division of Genetics, Indian Agricultural Research Institute, New Delhi; Institute of Agricultural Sciences, Banaras Hindu University, Varanasi; Indian Institute of Pulses Research, Kanpur; University of Agricultural Sciences, Dharwad; Panjabrao Deshmukh Krishi Vidyapeeth, Krishinagar, Akola from India participated in this consortium. This work was carried out with funding support from ICAR.

The consortium also used 'Asha' for assembling draft genome sequence data using long-sequence reads of 454 GS-FLX. A total of 0.51-Gb high-quality sequence data were generated and analysed. In brief, sequence analysis provided a total of 47,004 protein-coding genes which is quite similar to Varshney et al. (2012). Further, 1213 disease resistance/defence response genes were detected.

In summary, both of above-mentioned sequencing efforts have made the first pulse crop with genome sequence information. These efforts have created a 'Supermarket' of genes and molecular markers. However, we understand that number of genes predicted in pigeonpea are inflated and may reduce with better quality of draft genome assembly in near future. It will lead to the identification and manipulation of candidate genes or genomic regions associated with resistance or tolerance to biotic and abiotic stresses, yield contributing and other agronomic traits to enable pure line or hybrid breeding.

8.3 Deployment of Genome Sequence Information for Pigeonpea Improvement

Availability of genomic resources, such as draft genome sequence alone is not enough to improve the crop productivity. In fact, the genome sequence is starting point and one of the important tools to harness genetic diversity for the traits of interest to crop improvement. In addition to developing the varieties or parental lines of hybrids with enhanced resistance to sterility mosaic disease (SMD) and fusarium wilt (FW), the pigeonpea breeding community requires early maturing as well as photo-period insensitive lines so that crop production can be expanded to new niches such as sloping hills and fit into the new production systems with short-time windows. While resistance to pod borers (*Helicoveropa armigera* and *Maruca vitrata*) is another interesting trait, limited or non-availability of resistance to these insects in the cultivated gene pool does not allow the use of a molecular breeding approach for targeting these traits (at least for now).

Although breeders have been engaged in developing superior varieties, the genetic base of pigeonpea is limited due to minimal usage of diverse genotypes in breeding programs, and there have been only a few breeding programs across the world. Also, modern breeding tools such as molecular breeding could not be deployed in breeding due to non-availability of marker–trait information for traits of interest. In this direction, development of draft genome sequence has been considered as a milestone in pigeonpea research. As a result of this breakthrough, a significant amount of information has become available. These resources can be used as tools to harness the genetic diversity for crop improvement.

Just after decoding of pigeonpea genome sequence (Nov. 7, 2011), round table discussions were organized on 15 November, 2011 by Mr P. Basu, the then Secretary and Mr Mukesh Khullar, the then Joint Secretary, Department of Agriculture & Cooperation (DAC), Ministry of Agriculture, Government of India. In this meeting, representatives from different pigeonpea research institutes across India including the leaders of both genome sequencing projects—Dr Rajeev K. Varshney (IIPG/ICRISAT) and Dr NK Singh (NRCPB/ICAR)—and several other key stakeholders from India were present. In this important meeting, key issues related to pigeonpea improvement were identified. Subsequently,

ICRISAT in collaboration with University of Agricultural Sciences, ARS—Gulbarga, Karnataka and the Professor Jayashankar Telangana State Agricultural University (PJTSAU)-Hyderabad, Telangana—developed a road map for deploying genome sequence information for pigeonpea improvement. The consortium brought expertise in different disciplines, namely breeding, genomics, pathology, and wide hybridization. These efforts led to the development of the project 'Pigeonpea improvement using molecular breeding' with an aim to harness diversity present in the breeding material and germplasm collection. Subsequently, the proposal was approved by Ministry of Agriculture, Government of India and United States Agency for International Development (USAID)—India Mission.

The project was planned in three phases, phase I (generating basic information/material to initiate molecular breeding), phase II (molecular breeding, multi-location evaluation, development and extension in Asia and Africa) and phase III (Coordinated Research Project trials and extension in Asia and Africa) (Fig. 8.1). Phase I (for three years) was initiated in the year 2012 with the funding from USAID—India. Furthermore, research efforts were expanded for deploying genome sequence information for pigeonpea improvement through support from Department of Biotechnology, Government of India; CGIAR Research Program on Grain Legumes.

Significant progress has been made in different areas including developing new genetic stocks, utilization of genetic material for genome-wide profiling through re-sequencing, genotyping by sequencing, SNP arrays and establishing marker–trait association. Some of this progress has been presented in the sections below.

8.3.1 Developing New Genetic Stocks

Pigeonpea has a very strong crop improvement programme; however, pedigree analysis of released cultivars indicates that T-1 and T-190 were the most frequently used parents (Kumar et al. 2004). There has been limited use of genetic diversity from germplasm collection in breeding programs in majority of crops including pigeonpea (Upadhyaya et al. 2011). Above-mentioned points clearly explain reasons behind the low level of genetic diversity observed in cultivated gene pool. Therefore, it was planned to develop new genetic stock with broader genetic base and deploy family-based mapping approaches such as MAGIC and NAM to tackle complex traits precisely. These family-based approaches that play an intermediate role between classical bi-parental and natural populations have been developed in some other crops such as maize and *Arabidopsis* (McMullen et al. 2009; Kover et al. 2009).

Multi-parent advanced generation intercross (MAGIC): MAGIC could be used for high-resolution mapping and generating new breeding material with combined features of various important traits. Thus, MAGIC directly enriches breeding efforts in two ways: first, by revealing best alleles underlying a given trait and second, by their enrichment into a single genotype that can be easily recognized. MAGIC populations have been developed in *Arabidopsis* (Kover et al. 2009), rice (Bandillo et al. 2009, 2013) and wheat (Huang et al. 2012). Therefore, development of MAGIC population in pigeonpea through reshuffling of the genome to enhance the genetic base and to identify the marker traits associations was initiated. A total of eight diverse founder parents (four elite breeding lines and four landraces) with desirable features were selected for the development of MAGIC population (Table 8.1). Half diallel crossing approach (28, two-way F_1s) followed by funnel-based mating design (14, four-way and 7, eight-way F_1s) is being utilized for the development of at least 1000 MAGIC lines.

Nested association mapping (NAM): The NAM approach involves several populations that have one common parent, with the other parental genotype contrasting for traits of interest to the common parent genotype. Using this approach, available genetic diversity in the elite cultivars/breeding lines is utilized (McMullen

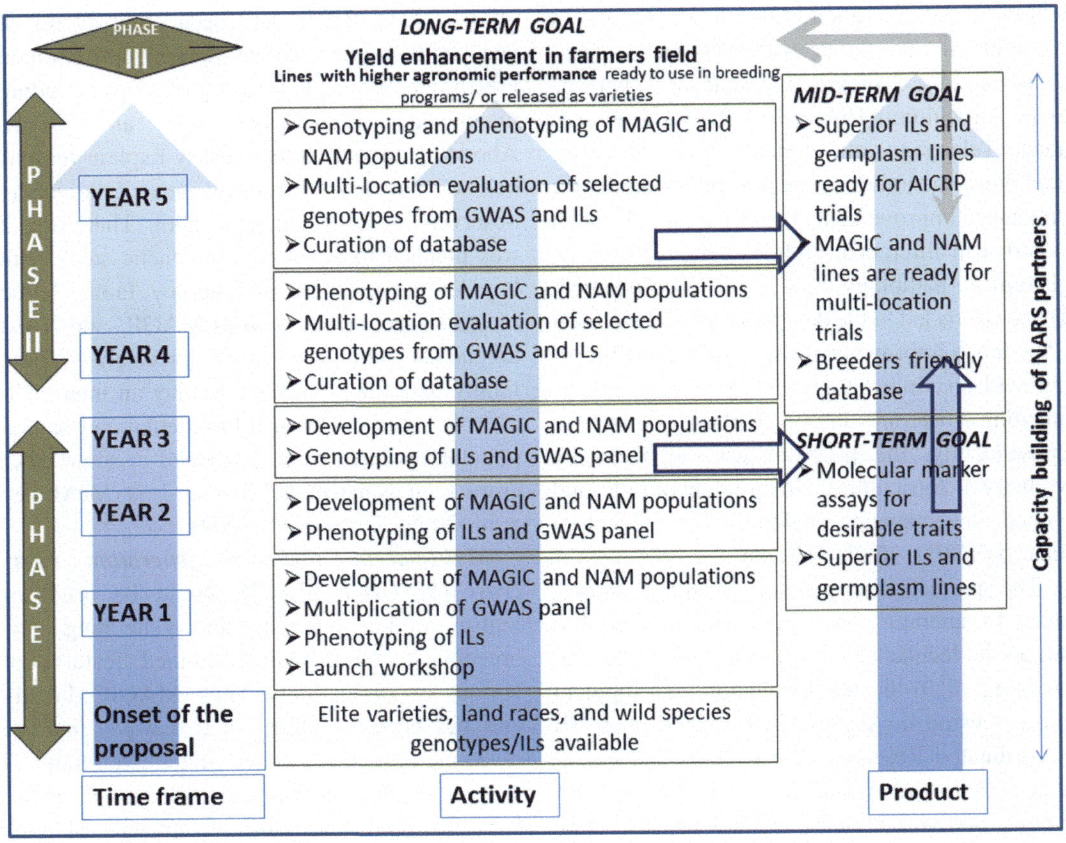

Fig. 8.1 A detailed plan for the deployment of genome sequence information for pigeonpea improvement in three phases, phase I (generating basic information/material to initiate molecular breeding), phase II (molecular breeding, multi-location evaluation, development and extension in Asia and Africa) and phase III (Coordinated Research Project trials and extension in Asia and Africa)

Table 8.1 List of parents and characteristic features used in the development of MAGIC population in pigeonpea

Lines	Features
ICP 7426	High pod numbers, medium duration
HPL 24	High protein content, medium duration, compact, susceptible to FW and resistant to SMD, inter-specific derivative
ICP 11605	Early flowering, germplasm line
ICP 14209	High number of pods, germplasm line
ICP 14486	Early flowering, germplasm line
ICP 5529	Medium duration, obcordate leaves, compact plant, poor yielding, modified flower
ICP 7035	Medium duration, SMD resistant to both Patancheru and Bangalore races, large purple seed, high sugar content
ICP 8863	Erect, mid-late, highly resistant to FW and susceptible to SMD, an extensively grown variety in Northern Karnataka and Maharashtra region of India, red seeded genotype

Table 8.2 List of parents and characteristic features used in the development of NAM population

Lines	Features
Nested parent	
Asha	Genome sequence available, leading variety, resistant to FW and SMD
Founder parent	
HPL 24	High protein content, medium duration, compact, susceptible to FW and resistant to SMD, inter-specific derivative
ICP 7035	Medium duration, SMD resistant to both Patancheru and Bangalore races, large purple seed, high sugar content
ICP 8863	Erect, mid-late, highly resistant to FW and susceptible to SMD, an extensively grown variety in Northern Karnataka and Maharashtra region of India, red seeded genotype
ICPL 87	Early duration, determinate, short, high combiner
ICPL 88039	Extra early maturity, indeterminate, good yield
ICPL 85063	Medium duration, indeterminate, good yield, more branching
MN 1	Super early, small seeded, determinate
ICP 28	Early maturity, local varieties
ICPL 85010	Early maturity, local varieties
ICP 7263	Determinate, long podded, white seeded

et al. 2009). Progenies coming from the different sets of bi-parental populations can be analysed using both linkage as well as association mapping approach. Therefore, there is a need to develop NAM populations for targeted traits. In the case of pigeonpea, a total of 10 crosses involving Asha as a common parent have been made at ICRISAT (Table 8.2).

Developed MAGIC and NAM populations in near future will be analysed together with genome-wide markers and trait phenotyping data. Such populations have potential to not only provide the trait-associated markers, but also provide new genetic combinations for pigeonpea breeding program, which can be proved as a game changer in elevating productivity and genetic diversity.

8.3.2 Utilizing Available Genetic Resources

As described in previous chapters of this book on genomic resources and germplasm characterization, a number of mapping populations ranging from F_2s (Varshney et al. 2010b), recombinant inbred lines (RILs), advanced back-cross mapping populations and specialized germplasm collections, namely composite collection (1000 lines, Upadhyaya et al. 2011), core collection (1290 lines, Reddy et al. 2005), mini-core collection (146 lines, Upadhyaya et al. 2006) have been developed. There is another set of germplasm collection comprising of 300 genotypes called 'reference set'. This 'reference set' represents 95% of SSR alleles present in the composite collection and also a reservoir of several traits of interests to the breeders, e.g. early flowering, high number of pods, high 100-seed weights and high seed yield per plant. To map the target traits for biotic stress resistance (FW and SMD), three bi-parental mapping populations comprising of 188 lines in each were developed and designated as Pigeonpea Recombinant Inbred Line (PRIL) populations. At present, there are three populations, namely PRIL_A (ICPB 2049 × ICPL 99050), PRIL_B (ICPL 20096 × ICPL 332) and PRIL_C (ICPL 20097 × ICPL 8863). Trait phenotyping of these mapping populations have been completed for two years at two to three locations. Based on the phenotypic datasets, FW and SMD resistance lines are being identified.

8.3.3 Whole-Genome Re-sequencing Initiatives

Initial investments in developing draft genome sequence information in any organism provide dividend in the form of re-sequencing projects. Where genome-wide profiling of target sets in a given species becomes cost-effective. Especially, in the present world of low-cost NGS, an opportunity arises to catalogue genome-wide variations in a number of individuals. NGS together with draft genome sequence has enabled identification of SNPs and Indels in efficient and high-throughput manner in a number of species (Lam et al. 2010; Xu et al. 2012). In the case of pigeonpea, re-sequencing of different germplasm sets has been initiated. These sets include parents of mapping populations (Kumar et al. 2016), reference set (Varshney et al. 2017), parental lines of hybrids (unpublished). These datasets will subsequently enable the pigeonpea researchers to deploy GAB and to overcome the bottleneck of limited genetic diversity information.

First generation HapMap: The first report on whole-genome re-sequencing (WGRS) in pigeonpea has been published very recently (Kumar et al. 2016). In this effort, WGRS was conducted using Illumina paired-end sequencing technology on a panel of 20 *Cajanus* lines. Selected lines represent crossing parentals of RILs, introgression lines (ILs), MAGIC and NAM. Across the 20 lines, a total of 157 Gb raw data with an effective mapping depth of $\sim$12X per genotype were generated. Subsequently, a total of 5.4 million variations including 4.6 million SNPs and 0.7 million Indels were identified across the lines. This study also provided 2598 copy number variations (CNVs) and 970 presence and absence variations (PAVs). Additionally, unique accession/genotype signatures were also detected through genome-wide analysis. The analysis clearly explained a narrow genetic base in cultivated gene pool and suggested use of new populations such as IL or AB populations, MAGIC, NAM to re-introduce adaptive diversity in pigeonpea breeding.

Reference set: To catalogue the sequence diversity, we have re-sequenced the genomes of 292 lines from a total of 300 pigeonpea lines in reference set on HiSeq 2500 platform. This set is comprised of breeding lines, landraces and wild species lines. WGRS yielded 2.15 Tb of sequence data with the coverage ranging from 5 to 12X. The WGRS data have been used to identify small variations (SNPs and Indels) and large structural variations (CNVs and PAVs). Re-sequencing data have provided variation counts around 17 million across 292 *Cajanus* lines (Varshney et al. 2017). This generated sequence data have been used to understand genetic relationships among *Cajanus* lines, targets of domestication and human selection genetic sweeps and associations between genomic regions with agronomic important traits.

Parental lines of hybrids: In order to define heterotic pools in pigeonpea, a set of 104 parental lines (cytoplasmic male sterile, maintainers and restorers) have been re-sequenced. WGRS yielded 511 GB sequence data with the coverage ranging from 5 to 10X. A total of 3.4 million SNPs could be identified across 104 lines (unpublished). In parallel, it has been planned to develop testcrosses from the parental lines. F_1 hybrids along with parental lines will be phenotyped for yield-related traits. The combined analysis with genome-wide variations and phenotypic data would provide clues on hybrid performance that will be helpful in accelerating hybrid improvement.

8.3.4 Sequencing-Based Genotyping Approaches

Pigeonpea draft genome sequence has provided opportunities to deploy sequencing-based genotyping approaches such as GBS and QTL-seq to generate high-density linkage maps and marker identification. GBS provides ease of discovery and genotyping of markers in a single-step process. GBS technique is now being used for genotyping PRILs (_A, _B and _C), multiple F_2 populations and ILs for developing high-density pigeonpea genetic maps and subsequently for identification of traits-associated markers (Saxena et al. 2017a, b). Another approach of marker detection combines bulked segregant analysis (BSA) and WGRS has been deployed in pigeonpea for identification of genomic region(s) responsible for FW and SMD resistance (Singh et al. 2016). In brief, R-bulk (resistant bulk) and S-bulk (susceptible bulk) along with the resistant parent were re-sequenced for identification of candidate genomic region(s)/genes. Detailed analysis of sequence data identified association of four candidate nsSNPs in four genes with FW resistance and four candidate nsSNPs in three genes with SMD resistance (for detailed results please see Singh et al. 2016).

8.3.5 High-Density SNP Array (Axiom®*CajanusSNP* Array)

In order to use above-mentioned sequence variations identified through large-scale WGRS in pigeonpea, an array targeting ~56,000 SNPs has been developed using the Affymetrix platform. SNP probes were designed by screening ~2 million SNP loci extracted from the re-sequencing data of cultivated lines. A total of 56,512 SNPs were placed on array. The array contained evenly distributed markers in the genome. Initial genotyping showed that this array had high genotyping accuracy and could be used for different objectives (unpublished). Therefore, it has been planned to genotype pigeonpea composite collection (~1000 lines) and at least three different mapping populations segregating for important traits with above-mentioned 'Axiom®*CajanusSNP* Array'. This array will play an important role in genomics studies and molecular breeding.

8.4 Conclusion

Two draft genome assemblies have been generated in pigeonpea by collaborative sequencing initiatives. These draft genomes have provided information on genes and subsequently led the foundation of a number of re-sequencing projects and applications of novel methodologies. The information produced through re-sequencing projects is continuously adding up and will be surely helpful in understanding the crop evolution, domestication, genetic relationships, genetic control of genes, biological process in different metabolic pathways etc. Whereas novel approaches such as GBS and SNP arrays together with trait phenotyping data will provide candidate markers/genes associated with different important traits in pigeonpea. As a next step in pigeonpea genomics, there is a need of improvement in draft genome assembly where PACBio reads along with improved analysis algorithms will certainly help. It is anticipated that coming years will be witnessing usage of genome sequence information in developing superior lines of pigeonpea for both varietal and hybrid breeding.

Acknowledgements Authors are thankful to the United States Agency for International Development (USAID), Biotechnology Industry Partnership Programme (BIPP) and Department of Biotechnology of Government of India for supporting various activities related to pigeonpea genomics and molecular breeding projects at ICRISAT. This work has been undertaken as part of the CGIAR Research Program on Grain Legumes. ICRISAT is a member of CGIAR Consortium.

References

Bandillo N, Muyco PA, Caspillo C, Laza M, Sajise AG, Redoña E, Singh RK, Leung H (2009) Population development through multiparent advanced generation intercrosses (MAGIC) among diverse genotypes to facilitate gene discovery for various traits in rice. Philippine J Crop Sci 34

Bandillo N, Raghavan C, Muyco PA, Sevilla ALM, Lobina TI, Dilla-Ermita CJ, Wei C, Tung CW, McCouch S, Thomson M, Mauleon R, Singh RK, Gregorio G, Redoña E, Leung H (2013) Multi-parent advanced generation inter-cross (MAGIC) populations in rice: progress and potential for genetics research and breeding. Rice 6:1–15

Bohra A, Saxena RK, Gnanesh BN, Saxena KB, Byregowda M, Rathore A, KaviKishor PB, Cook DR, Varshney RK (2012) An intra-specific consensus genetic map of pigeonpea [*Cajanus cajan* (L.) Millspaugh] derived from six mapping populations. Theor Appl Genet 125:1325–1338

Gnanesh BN, Bohra A, Sharma M, Byregowda M, Pande S, Wesley V, Saxena RK, Saxena KB, KaviKishor PB, Varshney RK (2011) Genetic mapping and quantitative trait locus analysis of resistance to sterility mosaic disease in pigeonpea [*Cajanus cajan* (L.) Millsp.]. Field Crops Res 123:53–61

Huang BE, George WA, Forrest KL, Kilian A, Hayden JM, Morell KM, Cavanagh RC (2012) A multiparent advanced generation inter-cross population for genetic analysis in wheat. Plant Biotechnol J 10:826–839

Kover PX, Valdar W, Trakalo J, Scarcelli N, Ehrenreich IM, Purugganan MD et al (2009) A multiparent advanced generation inter-cross to fine-map quantitative traits in *Arabidopsis thaliana*. PLoS Genet 5: e1000551. doi:10.1371/journal.pgen.1000551

Kumar S, Gupta S, Chandra S, Singh BB (2004) How wide is genetic base of pulse crops? In: Ali M, Singh BB, Kumar Shiv and Vishwa Dhar (eds) Pulses in new perspective. Proceedings of the national symposium on crop diversification and natural resources management, 20–22 December 2003. Indian Society of Pulses Research and Development, Indian Institute of Pulses Research, Kanpur, pp 211–221

Kumar V, Khan AW, Saxena RK, Garg V, Varshney RK (2016) First generation hapmap in *Cajanus* spp. reveals untapped variations in parental lines of mapping populations. Plant Biotechnol J 14:1673–1681. doi:10.1111/pbi.12528

Kumawat G, Raje RS, Bhutani S, Pal JK, Mithra AS, Gaikwad K, Sharma TR, Singh NK (2012) Molecular mapping of QTLs for plant type and earliness traits in pigeonpea (*Cajanus cajan* L. Millsp.). BMC Genet 13:1

Lam HM, Xu X, Liu X, Chen W, Yang G, Wong FL, Li MW, He W, Qin N, Wang B, Li J, Jian M, Wang J, Shao G, Wang J, Sun SSM, Zhang G (2010) Resequencing of 31 wild and cultivated soybean genomes identifies patterns of genetic diversity and selection. Nat Genet 42:1053–1059

McMullen MD, Kresovich S, Villeda HS, Bradbury P, Li H et al (2009) Genetic properties of the maize nested association mapping population. Science 325:737–740

Reddy LJ, Upadhyaya HD, Gowda CLL, Reddy KN, Singh S (2005) Development of core collection in pigeon pea [*Cajanus cajan* (L) Millspaugh] using geographic and qualitative morphological descriptors. Genet Resour Crop Evol 52:1049–1056

Saxena RK, Singh VK, Kale SM, Tathineni R, Parupalli S, Kumar V, Garg V, Das RR, Sharma M, Yamini KN, Muniswamy S, Ghanta A, Rathore A, Sameer Kumar CV, Saxena KB, Kavi Kishor PB, Varshney RK (2017a) Construction of genotyping-by-sequencing based high-density genetic maps and QTL mapping for fusarium wilt resistance in pigeonpea. Sci Rep 7:1911

Saxena RK, Kale SM., Kumar V, Parupalli S, Joshi S, Singh VK, Garg V, Das RR, Sharma M, Yamini KN, Ghanta A, Rathore A, Sameer Kumar CV, Saxena KB, Varshney RK (2017b) Genotyping-by-sequencing of three mapping populations for identification of candidate genomic regions for resistance to sterility mosaic disease in pigeonpea. Sci Rep 7:1813

Singh NK, Gupta DK, Jayaswal PK, Mahato AK, Dutta S et al (2012) The first draft of the pigeonpea genome sequence. J Plant Biochem Biotechnol 21:98. doi:10.1007/s13562-011-0088-8

Singh VK, Khan AW, Saxena RK, Kumar V, Kale SM, Sinha P, Chitikineni A, Pazhamala LT, Garg V, Sharma M, Sameerkumar CV, Parupalli S, Suryanarayana V, Patil S, Muniswamy S, Ghanta A, Yamini KN, Dharmaraj PS, Varshney RK (2016) Next generation sequencing for identification of candidate genes for fusarium wilt and sterility mosaic disease in pigeonpea (*Cajanus cajan*). Plant Biotechnol J 14:1183–1194. doi:10.1111/pbi.12470

Upadhyaya HD, Gowda CLL, Buhariwalla HK, Crouch JH (2006) Efficient use of crop germplasm resources: identifying useful germplasm for crop improvement through core and min-core collections and molecular marker approaches. Plant Genet Resour 4:25–35

Upadhyaya HD, Reddy KN, Sharma S, Varshney RK, Bhattacharjee R, Singh S, Gowda CLL (2011) Pigeonpea composite collection and identification of germplasm for use in crop improvement programmes. Plant Genet Resour 9:97–108

Varshney RK, Graner A, Sorrells ME (2005) Genomics assisted breeding for crop improvement. Trends Plant Sci 10:621–630

Varshney RK, Hoisington DA, Tyagi AK (2006) Advances in cereal genomics and applications in crop breeding. Trends Biotechnol 24:490–499

Varshney RK, Hoisington DA, Nayak SN, Graner A (2010a) Molecular plant breeding: methodology and achievements. In: Somers D, Langridge P, Gustafson PJ (eds) Methods in molecular biology: plant genomics. The Humana Press, TotowaNJ, USA, pp 283–304

Varshney RK, Penmetsa RV, Dutta S, Kulwal PL, Saxena RK, Datta S et al (2010b) Pigeonpea genomics initiative (PGI): an international effort to improve crop productivity of pigeonpea (*Cajanus cajan* L.). Mol Breed 26:393–408

Varshney RK, Chen W, Li Y, Bharti AK, Saxena RK, Schlueter JA et al (2012) Draft genome sequence of pigeonpea (*Cajanus cajan*), an orphan legume crop of resource-poor farmers. Nat Biotechnol 30:83–89

Varshney RK, Gaur PM, Chamarthi SK, Krishnamurthy L, Tripathi S, Kashiwagi J, Samineni S, Singh VK, Thudi M, Jaganathan D (2013) Fast-track introgression of "QTL-hotspot" for root traits and other drought tolerance traits in JG 11, an elite and leading variety of chickpea. The Plant Genome. doi:10.3835/plantgenome2013.07.0022

Varshney RK, Pandey MK, Janila P, Nigam SN, Sudini H, Gowda MVC, Sriswathi M, Radhakrishnan T, Manohar SS, Nagesh P (2014) Marker-assisted introgression of a QTL region to improve rust resistance in three elite and popular varieties of peanut (*Arachis hypogaea* L.). Theor Appl Genet 127:1771–1781

Varshney RK, Saxena RK, Upadhyaya HD, Khan AW, Yu Y, Kim C, Rathore A, Kim D, Kim J, An S, Kumar V, Anuradha G, Yamini KN, Zhang W, Muniswamy S, Kim J-S, Penmetsa RV, Wettberg EV, Datta SK (2017) Whole-genome resequencing of 292 pigeonpea accessions identifies genomic regions associated with domestication and agronomic traits. Nat Genet 49:1082–1088

Xu X, Liu X, Ge S, Jensen JD, Hu FY, Li X, Dong Y, Gutenkunst RN, Fang L, Huang L, Li JX, He WM, Zhang GJ, Zheng XM, Zhang FM, Li YR, Yu C, Kristiansen K, Zhang XQ, Wang J, Wright M, McCouch S, Nielsen R, Wang J, Wang W (2012) Resequencing 50 accessions of cultivated and wild rice yields markers for identifying agronomically important genes. Nat Biotechnol 3:105–157

9 Sequencing Pigeonpea Genome

Vikas K. Singh, Rachit K. Saxena and Rajeev K. Varshney

Abstract

Availability of draft genome has brought quantum jump in pigeonpea status and facilitated to move it to the league of genomic resource rich crops. It is important to mention that pigeonpea became the first orphan and non-industrial grain legume in 2012 to have the draft genome sequence. An elite pigeonpea genotype Asha (ICPL 87119) was used to develop the draft genome in two different sequencing efforts. The pigeonpea genome sequence effort led by International Crops Research Institute for the Semi-Arid Tropics (ICRISAT) used Illumina Genome Analyzer and HiSeq 2000 NGS platform, and a total of 237.2 Gb of sequence was generated. *De-novo* genome assembly combined with Sanger-based bacterial artificial chromosome end sequences and a genetic map was used to assemble raw reads into scaffolds representing 72.7% (605.78 Mb) of the 833.07 Mb pigeonpea genome. Genome analysis predicted 48,680 genes with an average transcript length of 2348 bp, coding-sequence size of 959.35 bp and 3.59 exons per gene. Analysis of genome assembly for repetitive DNA showed presence of transposable elements (TEs) in 49.61% of assembled genome. The pigeonpea genome sequencing led by National Research Centre on Plant Biotechnology (NRCPB) used 454 GS-FLX sequencing chemistry, with mean read lengths of >550 bp and >10-fold genome coverage, was used to assemble ~511 Mb sequence data. In this study, 47,004 protein-coding genes were predicted. This study also reported 1213 disease resistance/defense

V.K. Singh · R.K. Saxena · R.K. Varshney (✉)
International Crops Research Institute for the Semi-Arid Tropics, Patancheru 502324, India
e-mail: r.k.varshney@cgiar.org

R.K. Varshney et al. (eds.), *The Pigeonpea Genome*,
Compendium of Plant Genomes, DOI 10.1007/978-3-319-63797-6_9

response genes and 152 abiotic stress tolerance genes. The available pigeonpea draft genome information is expected to facilitate genomics-assisted breeding for the targeted traits that could improve food security in many developing countries.

9.1 Introduction

In continuation to the previous chapter in the book entitled "Whole-genome sequencing of pigeonpea: requirement, background history, current status and future prospects for crop improvement," we would like to present a focused chapter on de novo sequencing the pigeonpea genome. In the previous chapter, we have discussed about the background history of two genome sequencing efforts by Varshney et al. (2012) and Singh et al. (2012). However, in this short chapter, we present detailed comparisons between above-mentioned de novo sequencing projects in terms of (i) sequencing data, (ii) draft genome assemblies statistics, (iii) repetitive sequences in genome, (iv) gene annotation, (v) genome duplication and synteny with sequenced plant genomes, and (vi) novel marker and genes repertoire. Further, we have also presented utility of pigeonpea genome sequence by providing one example in soybean for crop improvement.

9.2 Sequencing Data

Genome sequencing of pigeonpea was undertaken in two different studies. Whole-genome shotgun sequencing strategy was used in both the studies. Illumina Genome Analyzer and Hiseq 2000 Sequencing System was used by Varshney et al. (2012), and 454 GS-FLX Phase D platform was used by Singh et al. (2012). Both of these next generation sequencing platforms have their advantages and disadvantages (Luo et al. 2012). In the case of Varshney et al. (2012), a total of 22 paired-end sequencing libraries with insert sizes of about 180 base pairs (bp), 250 bp, 350 bp, 500 bp, 800 bp, 2 kb, 5 kb, 10 kb, and 20 kb were used for sequencing on Illumina platforms. In the case of Singh et al. (2012), GS-FLX sequencing was undertaken on 20-kb-long fragments sequencing library. In these studies, 237.2 Gb and 10.1 Gb sequencing data were generated on Illumina sequencing and GS-FLX sequencing platform, respectively. Further, to reduce the effect of sequencing errors to the assemblies, a series of checking and filtering steps on reads generated were performed. After applying stringent criteria, only 130.7 Gb and 9.48 Gb data were used for developing draft genome assemblies by Varshney et al. (2012) and Singh et al. (2012), respectively.

9.3 Draft Genome Assemblies Statistics

In general, both studies had taken a series of steps to assemble the filtered/corrected sequencing reads. As the first step, raw sequencing reads were aligned to form contigs and then calculated the amount of shared PE relationships between each pair of contigs, and then constructed the scaffolds. Subsequently in Varshney et al. (2012), BAC-ends sequences (BESs) were used for mapping of scaffolds to obtain the super scaffolds, and a genetic map (*Cajanus cajan* ICP 28 × *C. scaraboides* ICPW 94) was used in developing the final scaffolds or pseudomolecules. A total of 137,542 and 59,681 scaffolds spanning 60,578 Mb and 5108 Mb genome assemblies were developed in Varshney et al. (2012) and Singh et al. (2012), respectively. The N50 of the assembly was 51,606 kb (scaffolds) in Varshney et al. 2012 and 4522 bp (contig) in Singh et al. (2012).

9.4 Repetitive Sequences in the Genome

In Varshney et al. (2012), repetitive DNA (excluding low-complexity sequences) was identified in 51.67% of the genome, most of which could not be associated with known transposable element (TE) families by the *de novo* repeat identification using RepeatModeler and homology analysis against the RepBase library. Majority of repetitive sequences were classified as retro-transposons (37.12%), whereas 8.77% of the transposable elements were DNA transposons. Long-terminal repeat elements, of which 22.81% are *Gypsy*-type elements and 12.04% are *Copia*-type element, were the most abundant. *De novo* analysis of RE were conducted using RepeatModeler software in the other study (Singh et al. 2012). As a result, a total of 1,127,729 REs in the were identified (63.95%) in the genome which covers a total of 326,671,068 bp sequences. Majority of the RE was retro-transposons (23.6%) (including Line: 1.03%; Copia 6.1%; and Gypsy 16.02%) and 2.99% was DNA transposons, whereas 66.2% was unclassified. Simple direct repeats and low-complexity repeats represented only 2.57% and 4.63% of the total RE, respectively.

9.5 Genome Annotation

Genome analysis combined with *de novo* gene prediction programs identified 48,680 pigeonpea genes (Varshney et al. 2012) (Table 9.1). The average transcript length was found to be of 2,348.70 bp, coding-sequence size of 959.35 bp and 3.59 exons per gene. *De novo* gene prediction supported majority of these predicted genes (99.6%). The annotation of the pigeonpea genome was found completed by observing 453 out of 458 (98.9%) KOGs within the pigeonpea gene set. The genes that have been predicted in pigeonpea genome are comparable to poplar (*Populus trichocarpa*), soybean (*Glycine max*), and *Medicago truncatula.* The average length of exon and intron in pigeonpea genome was found to be 267.39 bp and 536.89 bp, respectively, whereas the average number of exons per gene is 3.59. A total of 46,750 (96.04%) genes were found to be similar to entries in databases to tentatively assign gene functions and 1930 (3.96%) genes remain unannotated. Further, 862 microRNA (miRNA), 763 tRNA, 329 rRNA, and 363 small nuclear (snRNA) genes were identified in the pigeonpea genome set in addition to the protein-coding genes.

In the case of Singh et al. (2012), FGENESH software was used for gene prediction using 454 GS-FLX large sequence contigs containing ~511 Mb of high-quality sequence. A total of 59,515 genes were predicted with average gene size of 1170 bp. The gene with largest size was of 11,523 bp and the gene with smallest size of 501 bp. The average exon and intron sizes were 268 bp and 288 bp, respectively. The predicted 99.9% of the genes showed significant matches within the pigeonpea transcriptome database. A total of 47,004 protein-coding genes and 12,511 transposable elements related genes were reported. Additionally, 1213 disease resistance/defense response genes and 152 abiotic stress tolerance genes in the pigeonpea genome were also reported (Table 9.1).

After going in detail on these studies for last 2–3 years, we understand the number of genes predicted in both the genome assemblies mentioned above is an overestimate. This may be due to the quality of final genome assemblies which seemed to be fragmented. Improved version of assembly in near future may provide us the accurate number of genes in pigeonpea.

9.6 Genome Duplication and Synteny with Sequenced Plant Genomes

The synteny analysis in Varshney et al. (2012) revealed that pigeonpea diverged from soybean ~20–30 Myrs ago. In spite of this long period of divergence, high levels of synteny were observed between pigeonpea and soybean as well as between pigeonpea and the galegoid species *M. truncatula* and *Lotus japonicus*. Each pigeonpea chromosome showed extensive

Table 9.1 Comparative account on gene annotation in pigeonpea in two studies

Features	Varshney et al. (2012)	Singh et al. (2012)
Number of protein-coding genes	48,680	47,004
Number of gene models (non-TE containing)	40,071	34,493
Mean transcript length	2,348.70 bp	–
Mean coding-sequence length	959.35 bp	1170 bp
Mean number of exons per gene	3.59	4.90
Mean exon length	267.39 bp	268 bp
Mean intron length	536.89 bp	288 bp
Number of genes annotated	46,750 (96.04%)	–
Number of genes unannotated	1930 (3.96%)	–
Number of miRNA genes	862	100
Number of rRNA genes	329	448
Number of tRNA genes	763	671
Number of snRNA genes	363	226

synteny with two or more than two chromosomes in soybean, likely due to the independent duplication event in soybean following divergence from pigeonpea. The close relationships between pigeonpea and soybean genomes were also detected in Singh et al. (2012). In this study, a total of 31,937 (67.94%) of the pigeonpea genes showed synteny with soybean genes, whereas 9067 genes were unique to pigeonpea.

9.7 Novel Marker and Genes Repertoire

The completion of the pigeonpea genome has made a significant contribution to the genomic resources available for pigeonpea through sequencing of the pigeonpea genome. In Varshney et al. (2012), a total of 309,052 simple sequence repeats (SSRs) and 28,104 novel single nucleotide polymorphisms (SNPs) were identified. Further, a detailed comparative analysis has identified 111 drought-responsive genes for drought tolerance, an important trait that can be transferred to other legume crops, whereas in Singh et al. (2012) study, 1,89,895 SSRs comprising of 100,373 mono-nucleotide, 49,325 di-nucleotide, 18,505 tri-nucleotide, 2217 tetra-nucleotide, 512 penta-nucleotides, 815 hexa-nucleotide, and 18,148 compound repeats were reported. A total of 437 SSRs were experimentally validated for PCR amplification and high rate of polymorphism among pigeonpea varieties were reported.

9.8 Application of Pigeonpea Genome Sequence

The genome sequence provided hope to the pigeonpea community to use the genome sequence to harness pigeonpea's genetic diversity at genome level and to identify the molecular markers and genes for targeted traits. Such information will allow researchers to develop superior varieties and parental lines of hybrids in pigeonpea. The genome sequence will also be useful in identifying germplasm lines or advanced breeding lines with a broader genetic base for future breeding programs. Modern genetics and breeding approaches such as genotyping by sequencing, marker-assisted recurrent selection, and genomic selection will now be possible in this crop to improve the efficiency of pigeonpea breeding. Genome sequence will be useful in utilizing gene sequences in genetic engineering approaches also. Several projects are underway at present to harness the full potential of pigeonpea genome for crop improvement (Varshney 2016).

It is also important to mention here that the pigeonpea genome sequence information has also been used for crop improvement in other species like soybean. A very first example has come recently where pigeonpea genome has been used to bring rust resistance in soybean (Kawashima et al. 2016). In soybean, Asian soybean rust (ASR) is one of the most economically important diseases. This disease can only treatable with use of fungicides. However, due to the emergence of fungicide resistance in pest, it becomes less effective. Moreover, there are no commercial soybean cultivars with durable resistance. Interestingly, a gene *CcRpp1* (*Cajanus cajan* Resistance against *Phakopsora pachyrhizi 1*) from pigeonpea has been found useful to confer resistance to ASR in soybean. By analyzing the pigeonpea genome sequence, an intracellular immune receptor from pigeonpea was identified and transferred into soybean shows that *CcRpp1* confers full resistance to ASR in soybean. This will be helpful in achieving a higher level of resistance, which might provide commercial control superior to current strategies. This study has clearly shown the importance of pigeonpea genome and opened new avenues for its use not only in pigeonpea but also in other crops species (primarily closely related legumes) for crop improvement.

Acknowledgements Authors are thankful to the United States Agency for International Development (USAID), Biotechnology Industry Partnership Programme (BIPP) and Department of Biotechnology of Government of India for supporting various activities related to pigeonpea genomics and molecular breeding projects at ICRISAT. This work has been undertaken as part of the CGIAR Research Program on Grain Legumes. ICRISAT is a member of CGIAR Consortium.

References

Kawashima CG, Guimarães GA, Nogueira SR, MacLean D, Cook DR, Steuernagel B, Baek J, Bouyioukos C, do VA Melo B, Tristão G, de Oliveira JC, Rauscher G, Mittal S, Panichelli L, Bacot K, Johnson E, Iyer G, Tabor G, Wulff BHB, Ward E, Rairdan GJ, Broglie KE, Wu G, van Esse HP, Jones JDG, Brommonschenke SH (2016) A pigeonpea gene confers resistance to Asian soybean rust in soybean. Nat Biotechnol 34:661–665

Luo C, Tsementzi D, Kyrpides N, Read T, Konstantinidis KT (2012) Correction: direct comparisons of illumina vs. roche 454 sequencing technologies on the same microbial community dna sample. PLOS ONE 7 (3):10–1371

Singh NK, Gupta DK, Jayaswal PK, Mahato AK, Dutta S, Singh S, Bhutani S, Dogra V, Singh BP, Kumawat G, Pal JK, Pandit A, Singh A, Rawal H, Kumar A, Prashat RG, Khare A, Yadav R, Raje RS, Singh MN, Datta S, Fakrudin B, Wanjari KB, Kansal R, Dash PK, Jain PK, Bhattacharya R, Gaikwad K, Mohapatra T, Srinivasan R, Sharma TR (2012) The first draft of the pigeonpea genome sequence. J Plant Biochem Biotechnol 21:98–112

Varshney RK, Chen W, Li Y, Bharti AK, Saxena RK, Schlueter JA, Donoghue MTA, Azam S, Fan G, Whaley AM, Farmer AD, Sheridan J, Iwata A, Tuteja R, Penmetsa RV, Wu W, Upadhyaya HD, Yang SP, Shah T, Saxena KB, Michael T, McCombie WR, Yang B, Zhang G, Yang H, Wang J, Spillane C, Cook DR, May GD, Xu X, Jackson SA (2012) Draft genome sequence of pigeonpea *(Cajanus cajan)*, an orphan legume crop of resource-poor farmers. Nat Biotechnol 30:83–89

Varshney RK (2016) Exciting journey of 10 years from genomes to fields and markets: some success stories of genomics-assisted breeding in chickpea, pigeonpea and groundnut. Plant Sci 242:98–107

Future Prospects

10

Rajeev K. Varshney, Rachit K. Saxena
and Scott A. Jackson

Abstract

Pigeonpea with limited genetic diversity in the cultivated gene pool, long crop cycle, almost negligible public funding support to research as compared to other food crops remained an orphan crop. However, the development of extensive genetic stocks and genomics resources in recent years has made significant advances in pigeonpea research. Although genome sequence, genetic maps and a large set of markers allowed genome-wide identification of marker–traits associations and their deployment in breeding programs, there is a need for concerted community efforts to accelerate genetic gains in the crop breeding programs. This chapter proposes the use of a number of approaches that may be targeted by pigeonopea research community so that superior varieties or hybrids can be developed and disseminated to farmers in relatively short time. This will help to enhance the income of farmers as well as contributing to the food, nutritional and environmental sustainability in developing countries.

10.1 Introduction

More than 100 pigeonpea varieties and three hybrids (ICPH 2671, ICPH 2740 and ICPH 3762) have been released for cultivation (Singh et al. 2005; Saxena 2015), and thousands of germplasm accessions are present in the genebanks (Upadhyaya et al. 2006). However, the actual yield potential of pigeonpea has not been realized in farmers' fields (Mula and Saxena 2010). As discussed in many earlier chapters in this book, the yield levels remain stagnated during last seven decades. Moreover, the poor understanding of

R.K. Varshney (✉) · R.K. Saxena
International Crops Research Institute for the Semi-Arid Tropics (ICRISAT), Patancheru 502324, India
e-mail: r.k.varshney@cgiar.org

S.A. Jackson
The University of Georgia, College of Agricultural and Environmental Sciences, Athens, GA 30602, USA

R.K. Varshney et al. (eds.), *The Pigeonpea Genome*,
Compendium of Plant Genomes, DOI 10.1007/978-3-319-63797-6_10

genetics of many complex traits and limited concerted efforts in pigeonpea research and development has also been a contributing factor to develop the varieties and hybrids in relatively long time and with limited advantages over existing varieties/hybrids. Although hybrid technology has shown promise in elevating the yield levels in pigeonpea, its adoption has been limited to confined regions (Saxena et al. 2013). Very recently, new genomics approaches such as structural and functional genomics have started to enhance our understanding of genetic architecture of different traits in pigeonpea. Further, the advances in sequencing technologies have opened new opportunities for using a number of strategies to understand the complex traits and address the production constraints by developing new cultivars at a rapid pace. Therefore, in this chapter, we are highlighting futuristic approaches that can be used to enhance the genetic gains in pigeonpea breeding programs.

10.2 Genome Sequencing and Re-sequencing

Limited genetic diversity in crop species like pigeonpea poses threat to the crop to be more vulnerable to diseases, insect pests and climate change. Therefore, it is highly essential to continuously characterize and introduce novel genetic variations in crop breeding programs. This can be achieved by introducing mutants, landraces and wild species accessions related to the cultivated crop in breeding programs. However, linkage drag associated with favourable alleles in landraces and wild species accessions hindered their effective use in crop improvement (Sharma et al. 2013). The availability of next generation sequencing (NGS) and draft genome sequence in pigeonpea provides unique opportunities for exploring nucleotide-level diversity in cultivated, landraces and wild species accessions and its relationship to phenotypic diversity (Varshney et al. 2012). Re-sequencing of germplasm accessions will provide a better understanding of existing genetic diversity, associating gene(s) with phenotypes and exploiting natural genetic diversity to develop superior genotypes (Varshney et al. 2017). Few recent efforts as mentioned in the book chapters of re-sequencing pigeonpea reference set, hybrid parental lines and parental lines of mapping populations are underway. However, these efforts need to be enhanced exponentially and probably the entire germplasm collections stored in different genebanks should be sequenced. Ideally, we would like to see *de novo* genome assembly for each of the available *Cajanus* species and then re-sequencing data for all accessions. Further to use this information effectively, it is also important to have extensive phenotypic data on these collections so that appropriate marker–trait associations can be established.

10.3 Trait Mapping

We propose to undertake trait mapping in pigeonpea as mentioned in the following sections:

i. Prioritization of traits

   In order to enhance crop productivity, pigeonpea research community is always concerned for multiple traits related to yield and quality. Besides these traits, research is also focused on biotic and abiotic stresses to provide stability. It is important to prioritize traits as per the availability of human and financial resources in a given breeding program. Keeping this aspect in consideration, we have grouped target traits into three categories (Table 10.1). Category (Cat) 1 traits include immediate (5 years) need traits, Cat 2 traits are long-term (10 years) needs traits and Cat 3 are important traits but difficult to breed (>10 years). Therefore, high priority in implementation should be given to Cat 1 and 2 traits. For Cat 3 traits, consortium mode approach should be followed by bringing advanced research institutes and universities together.

ii. Rapid detection of markers associated with traits

Table 10.1 List of the traits on the basis of Cat 1 immediate (5 years) need, Cat 2 long-term (10 years) needs and Cat 3 important but difficult to breed (>10 years)

Trait	Immediate attention (Cat 1)	Long-term attention (Cat 2)	Difficult, but important (Cat 3)
Early maturity group	Super early[a]	Pod borer[c]	Photo-insensitivity[a]
	Large seed[a]	Phytophthora[d]	Races of phytophthora[d]
	High yield[a]	Water-logging[d]	
	Indeterminate[b]		
	Flowering time[a]		
	Wide adaptation[a]		
	Stable fertility restorers[b]		
	Primary branches[a]		
	Cleisto flower[a]		
Medium/late maturity group	Fusarium wilt resistance[b]	Races of fusarium wilt[d]	Photo-insensitivity[a]
	Sterility mosaic resistance[b]	Races of sterility mosaic[d]	Drought tolerance[a]
	High yield[a]	Phytophthora[d]	Races of phytophthora[d]
	Cleisto flower[a]	Rapid growth[d]	High protein[a]
	Flowering time[a]	Pod borer[c]	
	Hybridity test[b]	New CMS[a]	
	Diverse fertility restorers[a]		
	Primary/secondary branches[a]		
	Obcordate leaf shape[a]		

[a]Genomics interventions ongoing
[b]Genomics interventions have reached to advanced stages and providing inputs in regular breeding program
[c]Transgenic approaches required
[d]Target traits for implementing systematic genomics approaches along with other disciplines

To improve pigeonpea lines for "must have traits" as mentioned above through genomics-assisted breeding approaches, identification of traits-associated markers/genomic regions/quantitative trait loci (QTLs) or candidate genes are prerequisite. Once a marker (or candidate gene) associated with target traits is identified and validated; marker-assisted selection (MAS), marker-assisted backcrossing (MABC) or early generation screening (EGS) can be deployed for precise and rapid development of improved cultivars. Traditional QTL mapping approach involves identification of parental polymorphisms and genotyping the populations with polymorphic markers in time-consuming and resource intensive manner (Abe et al. 2012). Another common trait mapping approach is bulked segregant analysis (BSA) where marker screening on the extreme bulks and parents provides trait-associated markers (see Semagn et al. 2010). With the advantage of NGS technologies and draft genome sequence, rapid trait mapping has been performed for sterility mosaic disease (SMD) and fusarium wilt (FW) (Singh et al. 2015, 2017). These approaches depend on extreme bulks and NGS to even provide markers for qualitative and quantitative traits. We anticipate in future, heavy use of NGS-based BSA approaches for rapid and accurate trait mapping.

iii. High-resolution mapping

Other than bi-parental mapping populations, multi-parents mapping populations such as multi-parents advanced generation inter-cross (MAGIC) and nested association mapping (NAM) populations are being developed in pigeonpea for high-resolution mapping at ICRISAT. These family-based populations can be used for precise marker trait association following either individual or combination of genome-wide association studies and QTL mapping. MAGIC population in particular is useful for gene-trait association at higher resolution (Huang et al. 2012). On the other hand, NAM has advantage in identifying QTLs governing complex traits with higher phenotypic variation (McMullen et al. 2009). The selection of parental lines in developing MAGIC and NAM populations, however, plays an important role. RILs developed after this multi-dimensional crossing will have genetic architecture from a number of elite cultivars (crossing parents) and provides opportunity to evaluate them for must have traits. In summary, this multi-parental populations-based mapping approach will provide accurate means of tagging traits as well line with enhanced genetic diversity (in the case of MAGIC). In our opinion, efforts should also be directed towards utilization of the genetic variations and useful traits existing in many wild relative species of pigeonpea. In this direction, few introgression lines (ILs) have been developed at ICRISAT. However, there is still a scope to move towards backcross inbred lines (BILs), chromosome segment substitution lines (CSSLs), stepped aligned inbred lines recombinant strain (STAIRS) by using exotic/unused material.

10.4 Functional Genomics

After developing comprehensive transcriptome assemblies (Kudapa et al. 2012) and draft genome sequence (Varshney et al. 2012), there was a need to correlate and complement the genome information to the gene expression that is modulated in a temporal and spatial manner. In this direction, a very first step has been accomplished by developing a gene expression atlas for pigeonpea (Pazhamala et al. 2017). This will be helpful in studying the genes expressed in specialized tissues/organ system in pigeonpea. Further to harness the power of natural variation at tissue level in terms of protein and metabolites, the functional genomics should move towards the development of proteome and metabolome maps in pigeonpea. The protein and metabolic differences among tissues within a plant offer another source of variation that can be harnessed in the quest to understand gene function and tackling constraint related to crop improvement.

10.5 Next Generation Breeding

Majority of the present breeding programs mainly rely on phenotypic selection in standard breeding schemes. Few initiatives have been taken in pigeonpea improvement programs to take genomics inputs (marker-based purity testing in hybrids and parental lines; DNA fingerprinting, etc.) for enhancing the precision. Recently, GAB for introgression of SMD and FW resistance in elite varieties has also been started at ICRISAT. The availability of draft genome, re-sequencing data, NGS, bio-informatics advances and phenotyping platforms provide opportunities to take pigeonpea improvement program a step forward to move towards the next generation breeding. Recent drop in marker genotyping cost (lower cost per data point) will make next generation breeding possible in pigeonpea. These advances will enable breeders to select the most appropriate allele combinations for a number of traits (especially for must have traits) simultaneously at early generation and facilitating their introgression from landraces or wild species accessions into elite cultivars while avoiding the risk of linkage drag.

The use of genomic selection (GS) is becoming feasible in many crops (Zhao et al. 2015). GS is expected to help to design new genotypes based on genome-wide profiling in a cost effective and relatively fast way. In particular, pigeonpea hybrid breeding program, GS model can be designed for developing heterotic groups of parental lines. This will improve the chances of developing high-yielding better hybrids and parental lines. Once established, GS will eliminate the need for multi-location field testing at each generation and efforts in line × tester crossing. Though we understand that genetic resources, from elite cultivars to landraces and wild species accessions, will remain the foundation of pigeonpea improvement, we expect that in near future the use of NGS or high-throughput genotyping for EGS, GS, MABC and MAS will be accelerated. In parallel, new genome engineering approaches such as genome editing will also be very useful as and when causal nucleotide affecting the trait is identified.

Acknowledgements Authors are thankful to their colleagues and collaborators like KB Saxena, CV Sameerkumar of ICRISAT and Jochen Reif of IPK-Gatersleben, Germany, for discussions and suggestions that have helped to write this chapter.

References

Abe A, Kosugi S, Yoshida K, Natsume S, Takagi H, Kanzaki H, Matsumura H, Yoshida K, Mitsuoka C, Tamiru M (2012) Genome sequencing reveals agronomically important loci in rice using MutMap. Nat Biotechnol 30:174–178

Huang BE, George AW, Forrest KL, Kilian A, Hayden MJ, Morell MK, Cavanagh CR (2012) A multiparent advanced generation inter-cross population for genetic analysis in wheat. Plant Biotechnol J 10:826–839

Kudapa H, Bharti AK, Cannon SB, Farmer AD, Mulaosmanovic B, Kramer R, Bohra A, Weeks NT, Crow JA, Tuteja R, Shah T, Dutta S, Gupta DK, Singh A, Gaikwad K, Sharma TR, May GD, Singh NK, Varshney RK (2012) A comprehensive transcriptome assembly of pigeonpea (*Cajanus cajan* L.) using Sanger and second-generation sequencing platforms. Mol Plant 5:1020–1028

McMullen MD, Kresovich S, Villeda HS, Bradbury P, Li H, Sun Q, Flint-Garcia S, Thornsberry J, Acharya C, Bottoms C, Brown P, Browne C, Eller M, Guill K, Harjes C, Kroon D, Lepak N, Mitchell SE, Peterson B, Pressoir G, Romero S, Rosas MO, Salvo S, Yates H, Hanson M, Jones E, Smith S, Glaubitz JC, Goodman M, Ware D, Holland JB, Buckler ES (2009) Genetic properties of the maize nested association mapping population. Science 325:737–740

Mula MG, Saxena KB (2010) Lifting the level of awareness on pigeonpea—a global perspective. International Crops Research Institute for the Semi-Arid Tropics, Patancheru

Pazhamala LT, Purohit S, Saxena RK, Garg V, Krishnamurthy L, Verdier J, Varshney RK (2017) Gene expression atlas of pigeonpea and its application to gain insights on genes associated with pollen fertility implicated in seed formation. J Exp Bot 68:2037–2054

Saxena KB (2015) From concept to field: evolution of hybrid pigeonpea technology in India. Indian J Genet 75:279–293

Saxena KB, Kumar RV, Tikle AN, Saxena MK, Gautam VS, Rao SK, Khare DK, Chauhan YS, Saxena RK, Reddy BVS, Sharma D, Reddy LJ, Green JM, Faris DG, Nene YL, Mula M, Sultana R, Srivastava RK, Gowda CLL, Sawargaonkar SL, Varshney RK (2013) ICPH 2671—the world's first commercial food legume hybrid. Plant Breed 132:479–485

Semagn K, Bjornstad A, Xu Y (2010) The genetic dissection of quantitative traits in crops. Electron J Biotechnol 13: article 14 13(5): 16–17

Sharma S, Upadhyaya HD, Varshney RK, Gowda CLL (2013) Pre-breeding for diversification of primary gene pool and genetic enhancement of grain legumes. Front Plant Sci 4:309. doi:10.3389/fpls.2013.00309

Singh NB, Singh IP, Singh BB (2005) Pigeonpea breeding. Advances in pigeonpea research. Indian Institute of Pulses Research, Kanpur, India, pp 67–95

Singh VK, Khan AW, Saxena RK, Kumar V, Kale SM, Sinha P, Chitikineni A, Pazhamala LT, Garg V, Sharma M, Sameerkumar CV, Parupalli S, Suryanarayana V, Patil S, Muniswamy S, Ghanta A, Yamini KN, Dharmaraj PS, Varshney RK (2015) Next generation sequencing for identification of candidate genes for fusarium wilt and sterility mosaic disease in pigeonpea (*Cajanus cajan*). Plant Biotechnol J. doi:10.1111/pbi.12470

Singh VK, Khan AW, Saxena RK, Sinha P, Kale SM, Parupalli S, Kumar V, Chitikineni A, Vechalapu S, Sameer Kumar CV, Sharma M, Ghanat A, Yamini KN, Muniswamy S, Varshney RK (2017) Indel-seq: a fast-forward genetics approach for identification of trait-associated putative candidate genomic regions and its application in pigeonpea (*Cajanus cajan*). Plant Biotechnol J. doi:10.1111/pbi.12685

Upadhyaya HD, Gowda CLL, Buhariwalla HK, Crouch JH (2006) Efficient use of crop germplasm resources: identifying useful germplasm for crop improvement through core and mini-core collections and molecular marker approaches. Plant Genet Resour 4:25–35

Varshney RK, Chen W, Li Y, Bharti AK, Saxena RK, Schlueter JA, Donoghue MTA, Azam S, Fan G, Whaley AM, Farmer AD, Sheridan J, Iwata A, Tuteja R, Penmetsa RV, Wu W, Upadhyaya HD, Yang SP, Shah T, Saxena KB, Michael T, McCombie WR, Yang B, Zhang G, Yang H, Wang J, Spillane C, Cook DR, May GD, Xu X, Jackson SA (2012) Draft genome sequence of pigeonpea (*Cajanus cajan*), an orphan legume crop of resource-poor farmers. Nat Biotechnol 30:83–89

Varshney RK, Saxena RK, Upadhyaya HD, Khan AW, Yu Y, Kim C, Rathore A, Kim D, Kim J, An S, Kumar V, Anuradha G, Yamini KN, Zhang W, Muniswamy S, Kim JS, Penmetsa RV, von Wettberg E, Datta SK (2017) Whole-genome resequencing of 292 pigeonpea accessions identifies genomic regions associated with domestication and agronomic traits. Nat Genet 49:1082–1088

Zhao Y, Li Z, Liu G, Jiang Y, Maurer HP, Würschum T, Mock HP, Matros A, Ebmeyer E, Schachschneider R, Kazman E (2015) Genome-based establishment of a high-yielding heterotic pattern for hybrid wheat breeding. Proc Nat Acad Sci 112:15624–15629

The manufacturer's authorised representative in the EU is Springer Nature Customer Service Centre GmbH, Europaplatz 3, 69115 Heidelberg, Germany. If you have any concerns regarding our products, please contact ProductSafety@springernature.com

Printed and bound by CPI Group (UK) Ltd, Croydon, CR0 4YY
12/08/2026
02194544-0003